MW01050620

AN INTRODUCTION TO ELECTROCHEMICAL CORROSION TESTING FOR PRACTICING ENGINEERS AND SCIENTISTS

WILLIAM STEPHEN TAIT, PH.D.

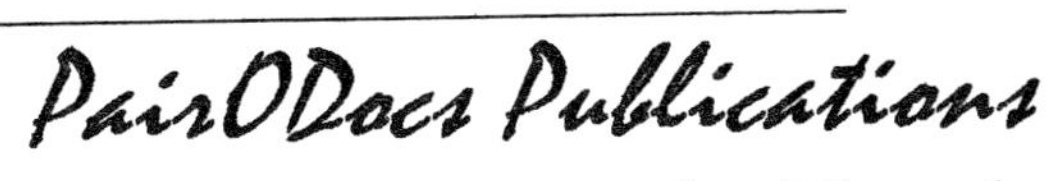

Racine, Wisconsin

ISBN 0-9660207-0-7

Printed in the United States of America

This book is dedicated to my wife Dr. Susan Whitworth Tait, and my daughters Rebecca Susan and Laura Stephanie

PREFACE

Electrochemical corrosion testing provides the means for predicting long term corrosion behavior and service lifetime of metallic structures, such as storage tanks, as well as monitoring of equipment to prevent catastrophic failure.

There are numerous technical papers that discuss electrochemical corrosion testing on a wide variety of metals and corrosive environments. The majority of these papers are written for peers in electrochemistry and corrosion engineering; few are written for those who use (or want to use) electrochemical corrosion testing without the benefit of having had graduate course work in electrochemistry or corrosion.

This book was written with the objective of providing engineers and scientists how-to-knowledge on the use of electrochemical corrosion testing to: a) solve corrosion problems, b) specify materials of construction for corrosive environments, c) determine service lifetime for a metallic structures without having the luxury of conducting long-term exposure tests, or d) monitor corrosion to prevent catastrophic failures from occurring.

It is not possible to discuss all types of corrosion data; nor will reading this book make one a corrosion expert. There will be situations in which one is confronted with data that are beyond the guidelines contained in this book. However, experience coupled with the principles covered in this book will provide you with the knowledge needed to make electrochemical corrosion testing a powerful tool for corrosion monitoring and prevention.

The text begins with a review of corrosion engineering and science basics, then applies these basics to electrochemical corrosion measurement methods. Mathematical equations are used only for purpose of instruction, leaving detailed derivations to the numerous corrosion textbooks that are available in book stores.

Chapters 1 through 3 provide background knowledge for why electrochemical corrosion measurement methods work and how to obtain the most reliable data from these measurements. While it is not necessary to read these chapters, You are encouraged to do so.

The topics in Chapters 4 through 6 follow an increasing direct current potential spectra sequence ranging from $\pm$ 10 to 20 mV from OCP used for linear polarization; to $\pm$ 250 mV from OCP for Tafel plots; -250 mV from OCP to approximately 1000 mV from OCP for potentiodynamic scanning; and finally from -250 mV from OCP, to 1000 mV from OCP and then back to OCP for cyclic polarization.

Chapters 7 and 8 discuss electrochemical impedance spectroscopy (EIS), an alternating current voltage corrosion measurement method.

I hope you enjoy reading this book and that it meets your objectives for learning more about electrochemical corrosion testing.

I would like to thank my editors, W. J. Eggers from EG&G Princeton Applied Research and E. Goldstien from G. E. Medical Systems, who helped to ensure that this book was both technically and grammatically correct. I would also like to thank the participants in my various continuing education courses whose questions provided much appreciated feedback on clarity (or the need for) of discussions in this book.

W. Stephen Tait, Ph.D. (October, 1994)

AN INTRODUCTION TO ELECTROCHEMICAL CORROSION TESTING FOR PRACTICING ENGINEERS AND SCIENTISTS

CONTENTS

CHAPTER 5

TAFEL PLOT CORROSION MEASUREMENT

CHAPTER 6

WIDE POTENTIAL RANGE DIRECT CURRENT POLARIZATIONS

CHAPTER 7

ELECTROCHEMICAL IMPEDANCE SPECTROSCOPY FUNDAMENTALS

CHAPTER 8

ANALYZING AND INTERPRETING EIS SPECTRA

LIST OF FIGURES

CHAPTER 1

Principles of Electrochemical Corrosion Important to Electrochemical Corrosion Testing

Objectives

After completing this Chapter, you will understand:

- why metallic corrosion is an electrochemical reaction
- the structure of the metal-electrolyte interface (electrical double layer), and how it affects metallic corrosion
- the relationships between electrical double layer (EDL) chemical composition, voltage, and electrical current
- how applied potentials can alter the chemical composition of the EDL
- how EDL composition changes relate to electrochemical corrosion measurements
- what types of electrochemical methods can be used to measure/study different types of corrosion

1.1 Introduction

Numerous technical papers have been written on electrochemical corrosion testing for a wide variety of metals and corrosive environments. The majority of these papers are written for experienced electrochemists and corrosion engineers with few written for engineers and scientist who are beginning to use electrochemical corrosion testing, without the benefit of having had graduate course work in corrosion engineering and/or electrochemistry.

This book does not discuss all possible electrochemical corrosion test methods, or interpretation of all types of corrosion data; nor will reading it make one a corrosion expert. This book is intended to provide background knowledge needed to allow practicing engineers and scientists to a) use electrochemical corrosion data to solve industrial corrosion problems, and b) understand and use information from technical corrosion journals to help interpret more complex electrochemical corrosion data than that covered in this book.

Chapters 1 through 3 provide the background knowledge for why electrochemical corrosion measurement methods work, and how to obtain the most reliable data from these measurements. Chapters 4 through 6 discuss the most commonly used direct current measurement methods, and Chapters 7 and 8 discuss electrochemical impedance spectroscopy.

1.2 Why use electronic equipment to measure corrosion?

The sensitivity of modern electronics allows measurement of corrosion long before either metal loss can be detected (by an analytical balance) or enough corrosion product accumulates on a metal surface so that it can be observed by the unaided eye. For example; a $3x10^{-9}$ Amp/cm^2 corrosion current density would cause approximately $1.1x10^{-5}$ grams of iron loss after 5 months, and would look like a tiny orange-brown, pin-point dot on the metal surface. In this case corrosion can be measured with electronic equipment (as an electrical current) much earlier than it can be seen or weighed.

Electrochemical corrosion measurements utilize the electrochemical nature of metallic corrosion. An external power source is used to apply a voltage, or range of voltages, to a metal specimen submerged in an electrolyte. The applied voltage, or voltage range, "pushes" the metal-electrolyte interface beyond its steady state conditions, causing a measurable electrical current to flow. Voltage and its corresponding current are independent and dependent variables (respectively), and their relationship is used to a) determine metallic corrosion behavior, and/or b) estimate corrosion resistance or impedance.

1.3 Corrosion Terminology

There is a glossary of corrosion science and testing terms at the end of this book. However, definitions of terms, most often, used in this chapter will be given at this point.

Corrosion current density

Corrosion current density is a corrosion rate expressed as an electric current. Corrosion current densities typically have units of Amps/cm^2 (A/cm^2), but can also have units like microAmps/cm^2 (μA/cm^2), or nanoAmps/cm^2 (nA/cm^2).

Electrolyte:

An electrolyte (in this book) is water or aqueous solutions containing dissolved ions and/or gases, such as oxygen.

Electrochemically active species:

Electrochemically active species are ions or molecules (e.g., hydrogen ions or oxygen) that can be reduced by electrons.

Electrode:

An electrode is a metal submerged in an electrolyte.

Open Circuit Potential:

Open circuit potential (OCP) is the electrical potential difference between two metals (submerged in an electrolyte) when no electrical current flows between them.

Polarization:

Polarization occurs when an electrical current shifts an electrode potential from OCP.

Potential

Potential is the electrical voltage difference between two electrodes, typically a reference and test electrodes. **Potential is used interchangeably with voltage throughout this book.**

Reference electrode:

A reference electrode is a cell containing a metal submerged in a specific concentration of its ions, plus ions from an inert salt. The cell is typically a glass tube having a semipermeable membrane or porous plug on one end that allows inert salt ions to enter and exit the cell.

1.4 Corrosion chemistry

Metallic corrosion occurs when metal atoms are oxidized and subsequently leave the metal lattice as ions. Valence electrons associated with metal ions (previously atoms) are left behind in the metal, creating an excess of electrons at the metal surface. The oxidation of metal atoms to ions is referred to as an electrochemical reaction because it is a chemical reaction that involves generation and transfer of electrons to electrochemically active species (dissolved) in the electrolyte. The transfer of electrons enables electronic measurement and study of metallic corrosion.

The oxidation half reaction of the metal is referred to as the anodic reaction and areas on a metal surface where oxidation occurs are referred to as anodes. The reduction of electrolyte electrochemically active species is referred to as the cathodic reaction and areas where reduction occurs are referred to as cathodes. Anodes and cathodes can have atomic dimensions or may be large enough to be observed with the unaided eye. Anodes and cathodes can be separated by finite distances as long as negative and positive ions can move in the electrolyte toward the anodes and cathodes, respectively, to maintain the electrical charge neutrality of the metal and electrolyte. Both anodic and cathodic reactions must be present to initiate and sustain metallic corrosion.

Electrochemical corrosion reaction equations contain symbols for electrons, reacting elements (and/or ions and molecules), and ions or molecules produced by the corrosion reaction. For example, the corrosion of iron is represented by the following anodic electrochemical equation:

$$Fe^{o} \rightarrow Fe^{+2} + 2e^{-} \qquad [1.1]$$

Fe^{o} represents iron atoms at the metal surface, Fe^{+2} represents iron ions, and $2e^{-}$ represents the two electrons produced by the anodic reaction. Equation 1.1 is referred to as an anodic half reaction because free electrons are produced, and a corresponding equation for the reaction of these electrons with an electrochemically active species is missing.

It was previously stated that both an anodic and cathodic reaction must be present to initiate and sustain metallic corrosion. Consequently, a cathodic half reaction must be written to account for reaction (reduction) of some electrochemically active species with excess electrons. An example of a cathodic half reaction is the reduction of hydrogen ions:

$$2H^{+} + 2e^{-} \rightarrow H_2 \qquad [1.2]$$

The overall corrosion reaction is written as the summation of both half reactions:

$$Fe^{o} \rightarrow Fe^{+2} + 2e^{-} \qquad \text{(anodic)} \qquad [1.1]$$

$$2H^{+} + 2e^{-} \rightarrow H_2 \qquad \text{(cathodic)} \qquad [1.2]$$

$$Fe^{o} + 2H^{+} \rightarrow Fe^{+2} + H_2 \qquad \text{(overall)} \qquad [1.3]$$

Equation 1.3 reads: iron atoms are oxidized to iron ions; producing electrons that reduce hydrogen ions (originating from the electrolyte) at the metal surface, forming hydrogen molecules.

Examination of the overall corrosion Equation 1.3, and the corresponding anodic and cathodic equations (Equations 1.1 and 1.2), leads to the hypothesis that corrosion can be halted by preventing one of the half reactions from occurring, and/or removing electrochemically active species from the electrolyte. Unfortunately it is difficult (but not impossible) to prevent half reactions from occurring, and there is a long list of electrochemically active species that cause corrosion. Water, for example, is electrochemically active and can cause corrosion of many commercially important metals such as mild steel and commercially pure aluminum. A partial list of other electrochemically active species includes oxygen (O_2), carbon dioxide (CO_2) dissolved in water, inorganic acids such as hydrochloric acid (HCl) and hydrogen sulfide (H_2S), and strong organic acids.

A metal submerged in an electrolyte will be referred to either as an electrode, or test electrode. A test electrode typically has millions of anodic and cathodic sites per square centimeter on its surface in approximately equal numbers. The metal and electrolyte together are electrically neutral, and no measurable external current flows to or from an electrode in the absence of an external applied voltage.

Electrochemical testing utilizes an external power source to apply a voltage, or range of voltages, to force an imbalance between the number of anodic and cathodic sites. Thus causing electrons to flow in an attempt to re-establish charge neutrality (balance). Electrons flowing to or from the electrode are electronically counted at each applied voltage level, yielding a data set consisting of a voltage (or voltages) and its (their) corresponding electrical current (or currents). Chapters 4 through 8 discuss how current-potential (voltage) data are used to determine corrosion behavior, rates and mechanisms.

1.5 The metal-electrolyte interface (electrical double layer)

Metal ions leave their lattice (leaving their electrons behind in the metal) when a metal is submerged in an electrolyte. Water molecules surround (hydrate) metal ions as they leave the lattice and the hydrated ions are free to diffuse away from the metal. The negative charge (caused by excess electrons) on the metal surface attracts positively charged metal ions and a percentage of them remain near the surface, instead of diffusing into the bulk electrolyte. The water layer around ions prevents most of them from making direct contact with excess surface electrons and subsequently being reduced to metal atoms. Positive ions in the electrolyte are also attracted to the negatively charged metal surface. Consequently, the electrolyte layer adjacent to an electrode surface contains water molecules, ions from the metal and bulk electrolyte, and has a distinctly different chemical composition than the bulk electrolyte. The negatively charged surface of a metal and the adjacent electrolyte layer are collectively referred to as the electrical double layer (EDL).

There are several EDL models[1] and Figure 1.1 contains one example that envisions a layer of water molecules (represented by open circles) absorbed on the metal surface, with hydrated metal ions (represented by encircled Me^{+n}) next to the adsorbed water layer.[2] Excess electrons are represented by a row of minus signs. The adsorbed water layer on the metal surface and water hydration sheath surrounding metal ions prevent them from contacting excess surface electrons, resulting in separate planes of positive and negative charges.

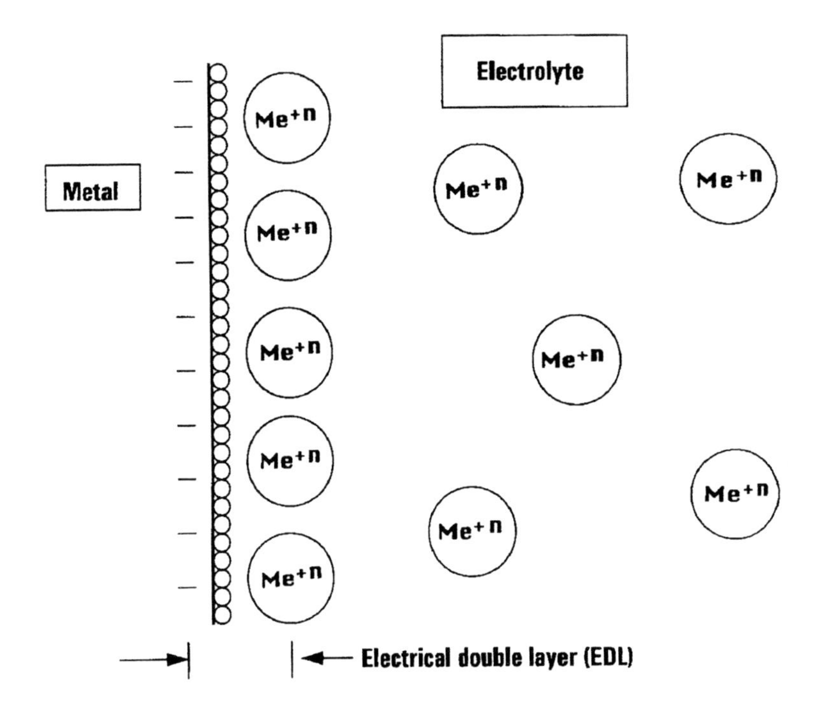

Figure 1.1. An equilibrium electrical double layer model for a metal submerged in an electrolyte

A corrosion reaction (such as the iron half reaction in Equation 1.1) can proceed in both the forward and reverse directions (i.e., both oxidation and reduction can occur), and equilibrium is achieved when forward and reverse reaction rates are equal. There is no net loss of metal atoms when an EDL 1.1 is at equilibrium, because metal atoms are oxidized and leave the lattice (as ions) at the same rate that metal ions are reduced and subsequently re-enter the lattice (as atoms). Thus the model in Figure 1.1 depicts a situation in which metal corrosion is at equilibrium.

Corroding metals, typically, are not at equilibrium and metal atoms continue to leave the metal lattice after the EDL has been established (i.e., the corrosion reaction proceeds in the forward direction as depicted in Equation 1.1). Corrosion is at steady state when it proceeds at a constant forward rate. Consequently, something must be added to the model depicted in Figure 1.1 to portray steady state corrosion like that for Equation 1.1; something that causes the EDL charge balance (charge neutrality) to become unbalanced and force metal atoms to continue leaving the metal lattice. What is missing from Figure 1.1 are electrochemically active species that can be reduced by the excess electrons, such as hydrogen ion reduction depicted in Equation 1.2. Electrochemically active species diffuse from the bulk electrolyte to the metal surface and discharge the EDL at the point on the metal surface where electrons are removed, causing more metal atoms to leave the lattice in an effort to re-establish original EDL conditions.

Figure 1.2 contains an EDL model for a corroding metal which includes electrochemically active species represented by solid circles.

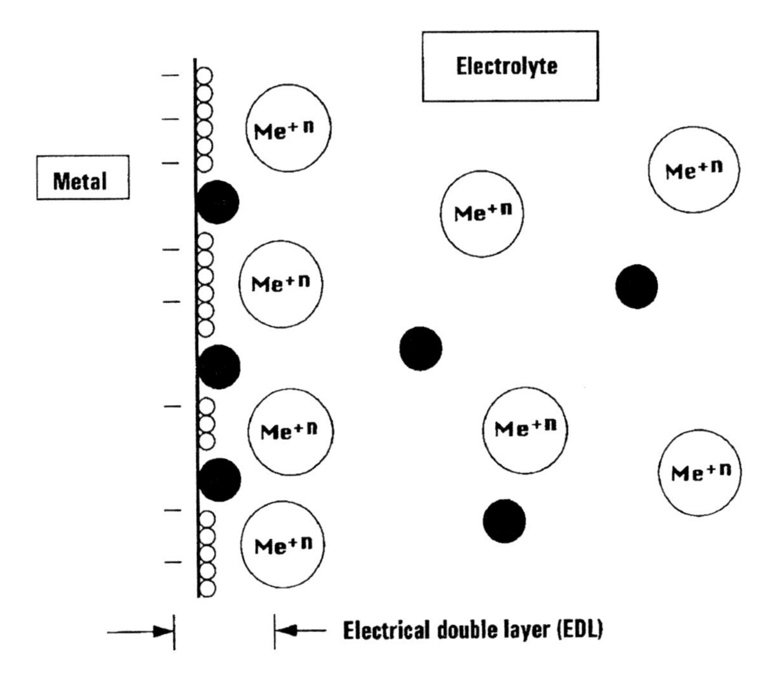

Figure 1.2. A steady state electrical double layer model for a corroding metal

An EDL is at steady state as long as electrochemically active species diffuse to the metal surface and remove electrons; and the corrosion reaction proceeds in the forward direction (metal atoms are continuously lost from the lattice) as shown in Equation 1.3.

Physically separating two oppositely charged planes produces an electrical capacitor. Consequently, charge (metal ions and electrons) separation gives an EDL capacitor-like behavior, and the level of EDL capacitance will be determined by the type of metal and associated electrolyte composition. A metal also resists transferring its excess electrons to electrochemically active species. Consequently, the EDL has both capacitive and resistive properties, and these properties are similar to those for a simple electrical circuit composed of a capacitor and resistor; such as shown in Figure 1.3. The capacitor is represented by the C_{EDL} symbol, and the resistor by the R_{ct} symbol.

The circuit in Figure 1.3 represents an equivalent model of the EDL, and is referred to as an equivalent electrical circuit model (EECM). Equivalent electrical circuit models will be used through out this book to help explain principles underlying electrochemical corrosion testing, and to model electrochemical impedance spectroscopy data (Chapters 7 and 8).

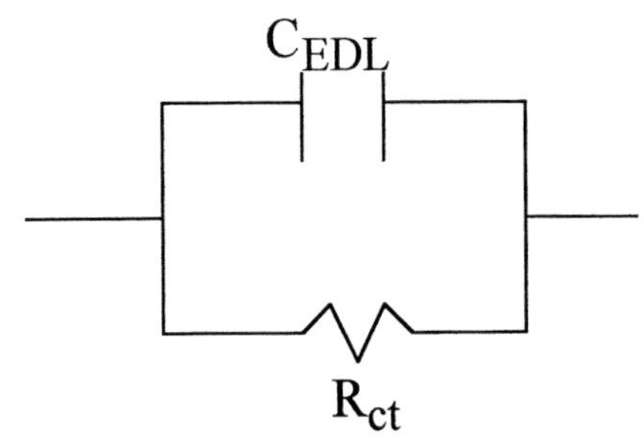

Figure 1.3. Simple electrical circuit having electrical properties similar to an EDL

Charge separation in an EDL also produces an electrical potential that can be measured as a difference between two metal electrodes, or a metal and reference electrode. Figure 1.4 depicts a test system schematic for measuring potential differences.

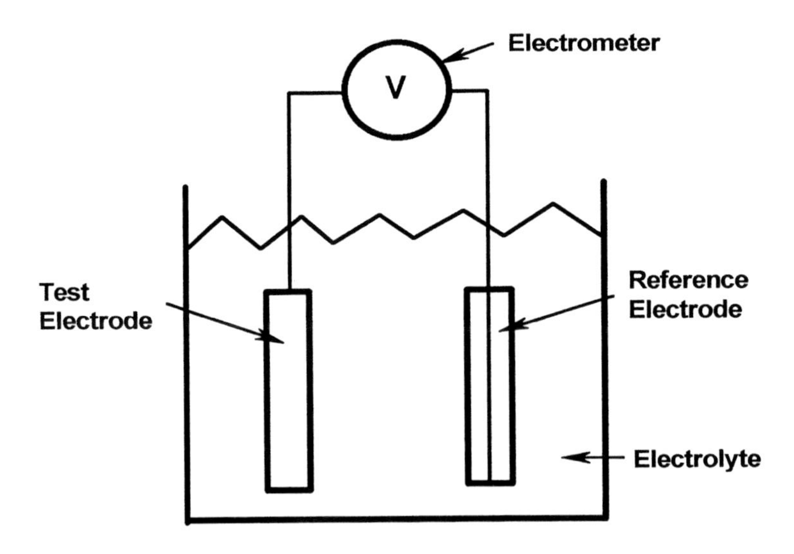

Figure 1.4. Schematic for measuring electrode potential (differences)

A test electrode (like that shown in Figure 1.4) is typically a coated or uncoated metal sample and the reference electrode can be either a true reference electrode (see terminology on page 3), or a pseudo-reference electrode fabricated from a metal that is corrosion resistant to the electrolyte. The potential difference is typically measured by electrically connecting two electrodes to an electrometer, which is a high impedance multimeter that can measure electrical voltage, resistance, and current.[3]

An electrode potential measured in absence of an applied potential is referred to as an open circuit potential (OCP).

The existence of a measurable electrical potential difference, and the involvement of electron transfer in the corrosion process, leads to speculation that there are probably relationships between EDL chemical composition, voltage and electrical current.

1.6 The Relationship between EDL Chemistry, Voltage and Electrical Current

Open circuit potential values change when EDL composition changes and applied voltages change EDL composition. The Nernst equation mathematically relates EDL composition to electrical potential:[4]

$$E = E^o - (RT/nF)\ln\{(a_{products})/(a_{reactants})\} \quad [1.4]$$

Where:

- **products** are chemical species (elements, ions or molecules) on the right side of an electrochemical equation such as Equations 1.1, 1.2 or 1.3
- **reactants** are chemical species on the left side of an electrochemical equation
- **a** is the chemical activity of products and reactants
- **E** is the measured potential in volts or millivolts
- $\mathbf{E^o}$ is the OCP when all activities in Equation 1.4 are equal to 1
- **R** is the ideal gas constant which is equal to 1.986 calories/mole ^oK
- **T** is temperature in degrees Kelvin(^oK)
- **n** is the number of electrons in the anodic half reaction (e.g., n = 2 in Equation 1.1)
- **F** is Faraday's constant which is equal to 96,500 coulombs/equivalent or 23,060 kcalories/volt
- The quantity RT/F is equal to 25.6 mV-equivalents at 298 ^oK (25^oC)

Chemical activity is equal to activity coefficient, γ, times concentration of a species in moles/liter; represented by an element symbol in brackets (e.g., [Fe^o]). Consequently, concentrations can be used in Equation 1.4 instead of activities and the Nernst equation can be rewritten as:

$$E = E^o - (RT/nF)\ln\{(\gamma_p[products])/(\gamma_r[reactants])\} \quad [1.5]$$

Where γ_p represents activity coefficients for products, and γ_r represents activity coefficients for reactants.

The Nernst equation for the iron anodic half reaction (Equation 1.1) is:

$$E = E^o - (RT/nF)\{\ln(\gamma_{+2}[Fe^{+2}])/(\gamma_o[Fe^o])\} \qquad [1.6]$$

where γ_{+2} is the activity coefficient for Fe^{+2}, and γ_o is the activity coefficient for iron atoms.

The activity of metal atoms (Fe^o in Equation 1.6) is considered to be equal to 1, hence the product, $\gamma_o[Fe^o]$, is equal to 1 and the Nernst equation for the iron half reaction becomes:

$$E = E^o - (RT/nF)\ln\{\gamma_{+2}[Fe^{+2}]\} \qquad [1.7]$$

The Nernst equation can also be written for the entire corrosion reaction. For example, the Nernst equation for corrosion of iron in an acidic electrolyte (Equation 1.3) is:

$$E = E^{o\prime} - (RT/nF)\ln\{(\gamma_{+2}[Fe^{+2}])/(\gamma_{H^+}[H^+]^2)\} \qquad [1.8]$$

Where γ_{H^+} is the activity coefficient for dissolved hydrogen ions and $E^{o\prime}$ is the potential when the activities of iron and hydrogen ions are both equal to 1. Water molecules and metallic iron are both omitted from Equation 1.8 because, by thermodynamic convention, elements have activity values equal to 1.

Equation 1.8 illustrates that the magnitude of a measured potential is determined by concentrations of both metal ions and electrochemically active species in the EDL. It follows that the magnitude of a measured potential will change when EDL chemical composition changes. Hence, OCP for a metal will change when the bulk electrolyte composition is altered in such a way that it causes EDL composition to change. Conversely, it also follows that changing the potential of a metal (using an external power source) will force the metal and electrolyte to supply ions/electrons and electrochemically active species, respectively, to the EDL and cause its chemical composition to change.

The Nernst equation does not express a relationship for electrical current. Electrical current is important to corrosion engineers and scientists because corrosion current can be converted into a rate of metal penetration by corrosion.

Ohms law states; that an electrical voltage is equal to the product of resistance times electrical current (V = IR). Hence it is reasonable to hypothesize that applying a potential to an electrode (in an electrolyte) will produce an electrical current, because an

EDL has both voltage and resistance. The Butler Volmer (*BV*) equation[5] relates electrical current to changes in metal potential caused by an external power source. The mathematical form of the *BV* equation is:

$$i = i_{corr} \{\exp\text{-}(\alpha nF\eta/RT) - \exp((1\text{-}\alpha)nF\eta/RT)\} \quad [1.9]$$

Where:

- R is the ideal gas constant, 1.986 calories/mole ^{o}K
- T is temperature in degrees Kelvin
- n is the number of electrons in the anodic half reaction
- F is Faraday's constant, 96,500 coulombs/equivalent or 23,060 calories/volt
- i is the external current density, in amps/cm^2, flowing to or from an electrode because of an applied potential
- i_{corr} is the corrosion current density, in amps/cm^2, that occurs when the electrode is at its OCP
- α is a coefficient having values that range from 0 to 1
- η is the test electrode overpotential and is the difference between electrode OCP and applied potential ($V_{applied}$ - OCP)
- F/RT is 0.039 mV^{-1} at 25° C

The exp-($\alpha nF\eta/RT$) term in Equation 1.9 is for cathodic current, and the exp($(1-\alpha)nF\eta/RT$) term is for anodic current. At OCP ($\eta = 0$) anodic and cathodic currents are equal and the external current, i, is zero (i.e., no current flows to or from the electrode).

Table 1.1

Comparing anodic and cathodic current levels at different overpotentials

Overpotential	Anodic current	Cathodic current
50mV	7.029 mA/cm^2	0.142 mA/cm^2
55mV	8.542 mA/cm^2	0.117 mA/cm^2
60mV	10.381 mA/cm^2	0.096 mA/cm^2
100mV	49.403 mA/cm^2	0.020 mA/cm^2

Table 1.1 contains values for anodic and cathodic currents (as current densities), calculated by the *BV* equation at different positive overpotentials. Notice that the cathodic current term in Equation 1.9 becomes insignificant when the overpotential is $\geq$ 50mV. For example, at 50 mV from OCP (i.e., η = 50mV) the external current, i, is approximately 98% anodic (7.03 mA/cm^2 anodic and 0.14 mA/cm^2 cathodic), and the external current is

essentially 100 % anodic at approximately 100 mV from OCP. Thus an electrode becomes (essentially) an anode when the overpotential is ≥ 50mV, allowing independent measurement and study of the anodic half reaction. Conversely, an electrode becomes a cathode when the overpotential is decreased below -50mV, allowing independent measurement and study of cathodic half reactions.

Making an electrode entirely anodic or cathodic produces external currents (to or from an electrode) which can be measured by electronic equipment. The importance (to electrochemical corrosion measurement) of being able to independently study half reactions will be fully discussed in Chapters 4 through 6.

1.7 Electrochemical corrosion curves, Evans diagrams, and mixed potential theory.

The Butler Volmer equation can be used to produce a plot of current as a function of voltage, such as that shown in Figure 1.5 for iron. The shape of this curve is similar to curves obtained from direct current polarization methods such as Tafel plots (discussed in Chapter 5), and beginning portions of both potentiodynamic scanning and cyclic polarization curves (discussed in Chapter 6). Log values of current are plotted along the X-axis, even though voltage is the independent variable. This type of plotting convention is a hold-over from older electrochemical research in which current was controlled and voltage was measured.

Figure 1.5 has anodic and cathodic branches. The iron anodic half reaction is written next to the anodic branch (top branch) of Figure 1.5, and the iron cathodic half reaction is written next to the cathodic branch (bottom branch). OCP for the corrosion reaction occurs at the curve inflection point, which is also the potential where no net current flows to or from the electrode (i.e., $\dot{i} = 0$ in Equation 1.9).

The corrosion current density ($\dot{i}_{corr}$ in Equation 1.9) can be estimated by extrapolating linear portions of the anodic and cathodic branches in Figure 1.5 to OCP, as shown by the solid lines. It will be shown in Chapter 5 how to convert corrosion current density to a corrosion rate.

Current-potential curves like Figure 1.5 can be simplified by showing only the solid lines, such as is done in Figure 1.6. Figure 1.6 is referred to as an Evans diagram, after U. R. Evans who first proposed this type of simplification.

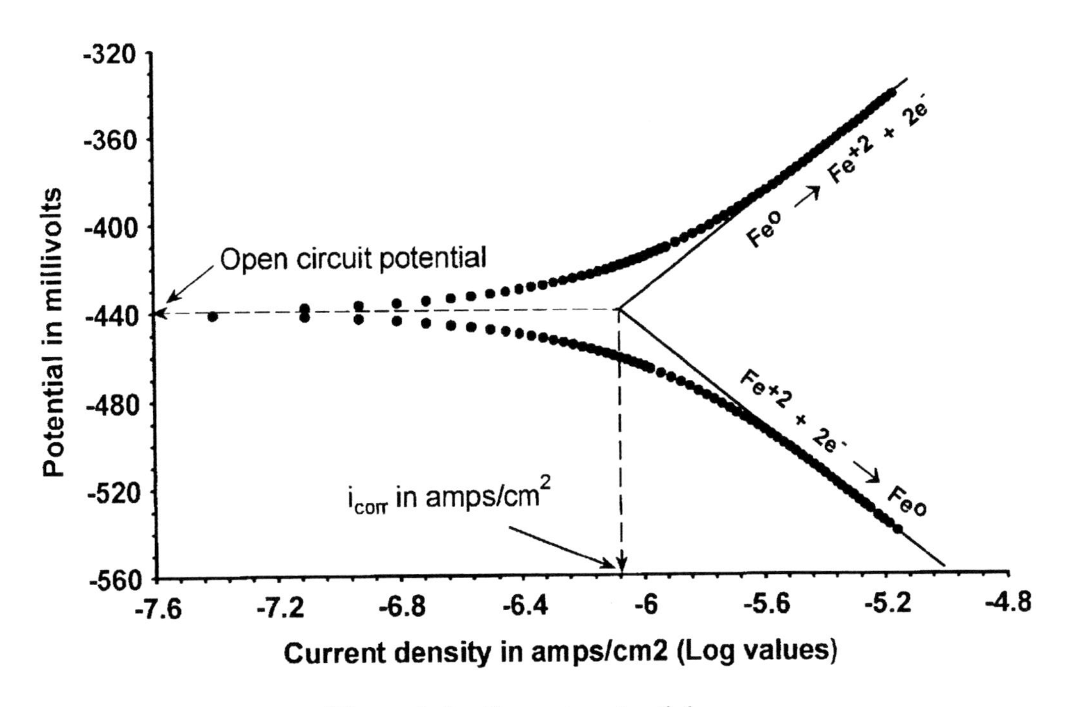

Figure 1.5. Current-potential curve

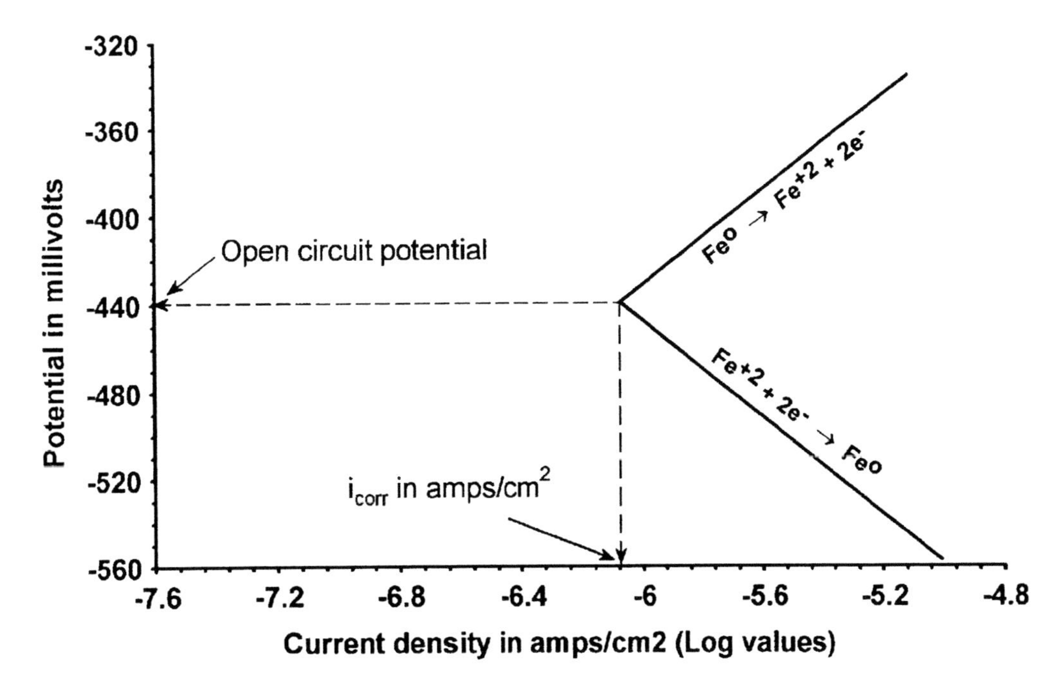

Figure 1.6. Evans diagram for iron

An Evans diagram can be used to model steady state corrosion for corrosion reactions like that for iron in acid (Equation 1.3 on page 4). Figure 1.7 contains an overlay of the Evans diagrams for hydrogen (Equation 1.2) and iron (Equation 1.1) and represents the corrosion or iron in acid (Equation 1.3). Hydrogen and iron corrosion current densities are labeled on the current axis and appropriate electrochemical half reactions are located along each branch.

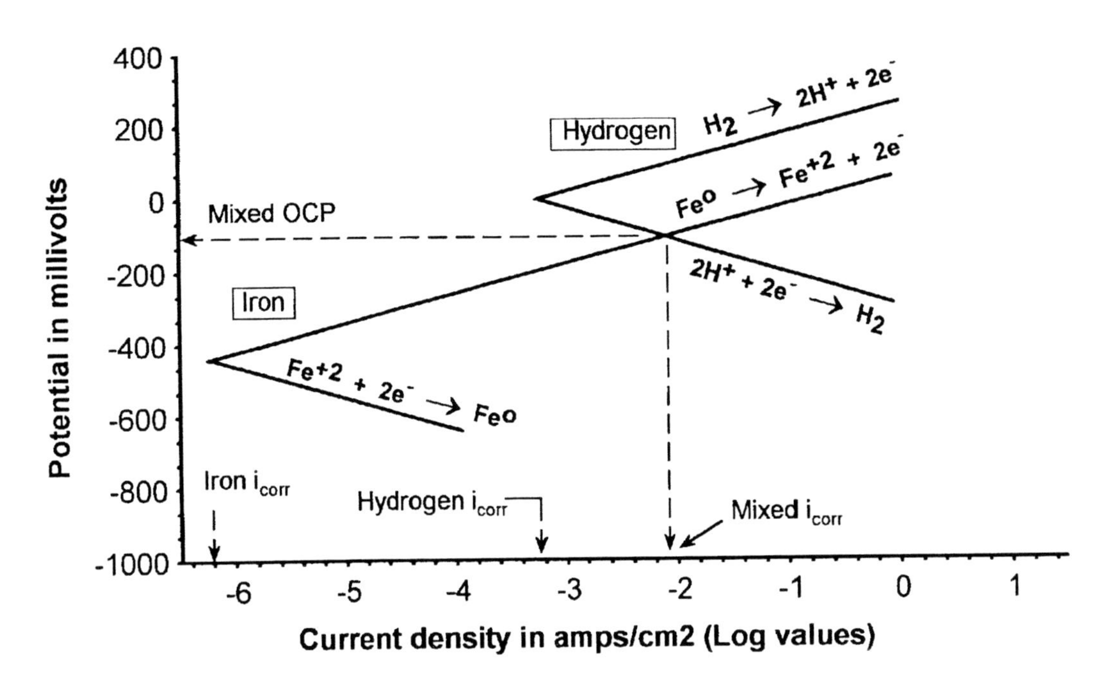

Figure 1.7: Evans diagram for corrosion of iron in an acid electrolyte

The Evans diagram in Figure 1.7 also illustrates that hydrogen ions are reduced at the surface of an iron electrode and cause corrosion to proceed in the forward direction. Figure 1.7 provides additional information on corrosion kinetics (i.e., corrosion rate) that is not in either Figure 1.2, or Equation 1.3. Notice in Figure 1.7 that the cathodic branch of the hydrogen diagram intersects the anodic branch of the iron diagram, and the OCP for the combined reaction (intersection point of the cathodic and anodic branches) is in-between OCP values for the individual reactions. Notice also that the corrosion current for the combined reaction (Equation 1.3) is higher than current densities for individual reactions; that is, hydrogen ion reduction causes iron corrosion to proceed at a higher rate (higher i_{corr}) than without hydrogen ions.

The concept of two electrochemical half reactions combining (as shown in Figure 1.7) to form a corrosion reaction, whose OCP and corrosion rate are different from that for the two half reactions, is referred to as mixed potential theory.

Evans diagrams will be used throughout this book with EECM's to help illustrate principles underlying electrochemical corrosion testing.

1.8 Types of corrosion that can be electrochemically measured

Corrosion can be separated into two types: a) uniform or general corrosion, and b) localized corrosion. General corrosion occurs uniformly over the entire metal surface. Localized corrosion occurs at small (dimensions ranging from 1 to 10 microns to several centimeters) discrete locations on a metal surface and is usually characterized by rapid, deep penetration through the metal. Localized corrosion can be several orders of magnitude faster than general corrosion.[6]

General corrosion rates can be measured electrochemically with linear polarization (discussed in Chapter 4), and Tafel plots (discussed in Chapter 5), and Electrochemical Impedance Spectroscopy (Chapters 7 and 8). General corrosion can also be characterized with potentiodynamic scanning (PDS) curves (discussed in Chapter 6).

Pitting and crevice localized corrosion can also be measured or characterized by electrochemical methods. Cyclic polarization (discussed in Chapter 6) can be used to characterize pitting and crevice corrosion, and in certain situations to estimate pitting corrosion rates.

1.9 Summary

The main topics covered in this Chapter were:

1. Metallic corrosion is a steady state electrochemical process in which metal atoms chemically transform to ions, and leave their valence electrons behind in the metal. Electrons left behind in the metal are transferred to electrochemically active species, causing corrosion to proceed in the anodic direction.
2. Chemical composition of a metal-electrolyte interface (electrical double layer) is significantly different from its metal surface and bulk electrolyte compositions. Charge separation in an electrical double layer (EDL) gives it electrical properties such as capacitance, resistance and voltage. EDL composition can be changed by either altering bulk electrolyte composition, or applying a voltage to an electrode.
3. Altering EDL composition in the absence of an applied voltage will change the corresponding electrode OCP. Applied voltages cause an electrical current to flow from an anode, or to a cathode.

4. Applied voltages and corresponding electrical currents are used to characterize, measure and study metallic corrosion.

1.10 References

1) J. O'M. Bockris & A. K. N. Reddy, Modern Electrochemistry, volume 2, pp. 623 - 843, Plenum Press, NY (1971)
2) J. Burgess, Metal Ions in Solution, pp. 19 - 23, distributed by John Wiley & Sons, NY (1978)
3) Low Level Measurements, third edition, p. 2, Keithley instruments, Cleveland Ohio (1984)
4) K. J. Vetter, Electrochemical Kinetics, p. 25, Academic Press, NY (1967)
5) A. J. Bard and L. R. Faulkner, Electrochemical Methods, p. 103, John Wiley & Sons, Inc., NY (1980)
6) R. R. Annand, H. M. Hilliard, and W. S. Tait, **STP641**, p. 41, American Society for Testing and Materials, PA (1978)

CHAPTER 2

Getting the Best Information from Electrochemical Corrosion Data

Objectives

After completing this Chapter, you will understand:

- how uncompensated solution resistance can affect electrochemical corrosion data
- how to correct electrochemical data for uncompensated solution resistance
- how electrode and cell geometries can affect electrochemical corrosion data
- how to design corrosion test cells and electrodes to get electrochemical corrosion data that represent actual corrosion behavior
- why it is important to use replicate measurements for each metal–environment variable
- the use of different statistical methods for analyzing electrochemical corrosion data

2.1 Introduction

Corrosion reactions (such as that in Equation 1.3 on page 4) are often complex heterogeneous reactions whose rates are typically determined by the interaction of a number of factors such as: a) the usual kinetic considerations (e.g., activation energy), b) electrolyte chemical composition, c) mass transfer between electrolyte and the metal surface, and d) various surface effects such as adsorption/desorption and surface roughness.[1] Interactions between these factors often make it difficult to reproduce electrochemical data and/or the exact conditions that cause a metal to corrode.

There are several experimental factors that must be considered when designing corrosion experiments (measurements) and test cells. Ignoring these factors can generate data that does not represent the actual corrosion behavior of the test metal being studied. For example, neglecting to compensate for solution resistance can produce electrochemical corrosion rates that underestimate actual rates and could result in a metal structure whose thickness is insufficient to prevent premature failure from thinning by corrosion.

Fortunately errors in electrochemical corrosion measurements can be minimized through control of several experimental (measurement) factors. This chapter discusses how to minimize inherent corrosion measurement errors and offers suggestions on how to cope with electrochemical data variability.

We begin with definitions for some of the more commonly used electrochemical corrosion terms in this chapter. The Reader may also want to review definitions of terms contained in Chapter 1 (pages 2 and 3), because these terms will also be used in this chapter.

2.2 More corrosion terminology

Current Path:

A current path is the path an ion follows as it moves under the influence of an electrical field between two electrodes in an electrolyte.

Electrical current:

Electrical current is the flow of electrons through electrically conductive solid materials such as metal electrodes and their corresponding electrical connections to a potentiostat.

Electrical current is measured in amps, which is the amount of electrical charge in Coulombs per second. Current can be converted to a metal corrosion rate, such as milligrams per year, using Faraday's law.[3]

Ionic current:

Ionic current is the flow of positive and negative ions under the influence of an electrical field between two electrodes in an electrolyte.

Ion mobility:

Ion mobility is the limiting velocity of an ion in a one volt electric field. Ion limiting velocity, in an electrolyte, is a function of ion concentration, the magnitude of charge on the ion, and frictional drag on the ion as it moves through the electrolyte.

Ion mobility is different than diffusion. Ion mobility occurs because of an electrical field between two electrodes and diffusion occurs because of a concentration difference between two areas in an electrolyte.

Potentiostat:

A potentiostat is an electronic device that is used to control the potential of a test electrode in an electrolyte. The magnitude of electrode potential change (polarization) is determined by the amount of electrical current supplied by a potentiostat.

Electrode potential is measured as the difference between itself and a reference electrode (see Chapter 1 definitions on page 3).

Solution resistance:

Solution resistance is the electrical resistance of an electrolyte.

2.3 The electrochemical experiment

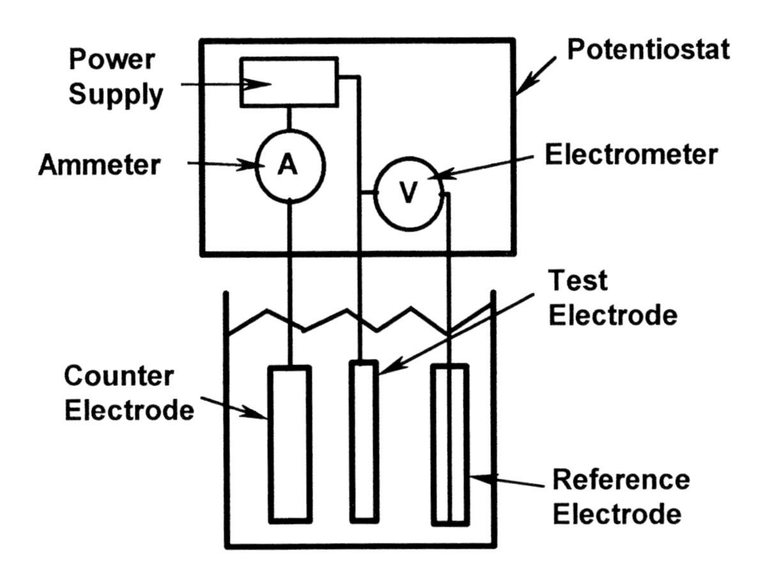

Figure 2.1. Three electrode test cell schematic

Figure 2.1 depicts a typical electrochemical corrosion test cell consisting of three electrodes submerged in an electrolyte. Electrical current from a potentiostat changes a test electrode potential from its open circuit potential (OCP), to a potential value that is determined by the magnitude of potentiostat current. Changing an electrode potential from its OCP is referred to as polarization, and Equation 1.11 on page 11 relates potentiostat current ("i" in Equation 1.11) to test electrode polarization (overpotential, η, in Equation 1.11).

Electrical current must be concurrently withdrawn from a test electrode when current is supplied (by a potentiostat) to a counter electrode (and vice versa), in order to maintain electronic equipment and electrode electrical neutrality.

Test electrode polarization is measured as a potential difference between reference and test electrodes. No electrical current flows between a potentiostat and reference electrode, so it remains at its OCP and provides a "fixed" reference point for corrosion measurements. The reference electrode is also used to provide feedback to the potentiostat, so that test electrode potential can be monitored and adjusted to a desired level. A thorough discussion of reference electrode properties and types is beyond the scope of this book. Consequently, the Reader is referred to references 4 and 5 cited at the end of this chapter for discussions on reference electrodes.

The relationship between a spectrum of electrode potentials and their corresponding electrical currents, or a range of voltage frequencies and corresponding impedance values, can be used to determine corrosion behavior such as:

- if a metal will passivate (not corrode)
- if pitting corrosion will occur
- if a coating will provide corrosion protection
- if metallic corrosion occurs under a coating

Potential–current or frequency–impedance data can also be used to determine corrosion parameters such as:

- electrical double layer capacitance
- corrosion rates
- corrosion resistance
- organic coating capacitance and pore resistance

2.4 What happens when current flows between test and counter electrodes?

Test electrode polarization is controlled by a potentiostat supplying electrons to either the counter or test electrodes, much like the number of cars/hour exiting a highway is controlled by the number of cars/hour entering the highway.

But everyone knows that the number of cars/hour entering a highway will not always control cars/hour exiting. There are a number of other factors such as number of open lanes, weather, road conditions, accidents, slow automobiles, and number of automobiles on a highway, that can also affect how many cars exit per hour. Similarly, there are also many factors other than potentiostat current that can control the magnitude of electrode polarization. Examples of these factors are: a) test electrode chemical composition b) test and counter electrode surface condition, c) test and counter electrode geometrical shapes, d) counter electrode size (area), e) electrical double layer chemical composition, and f) electrolyte chemical composition,.

Electrode polarization causes certain processes to occur in an electrolyte that can limit electrical current during electrochemical corrosion measurements, much like rush-hour traffic or road conditions can create situations that limit the number of cars/hour exiting a highway. Figure 2.2 contains a test cell schematic in which the reference electrode is excluded to allow space to illustrate various processes that occur during polarization.

Ions (imagine cars on a highway) respond to electrode polarization (rush-hour traffic) by moving between counter and test electrodes (entrance and exit ramps, respectively) in order to maintain electrical neutrality of the electrodes and electrolyte. Electrochemically active species also move to the counter electrode and react with electrons supplied by the potentiostat. The potentiostat in Figure 2.2 supplies electrons to the counter electrode, causing positive ions (cations) to move toward the counter electrode. The potentiostat withdraws electrons from the test electrode and negative ions (anions) move toward the test electrode.

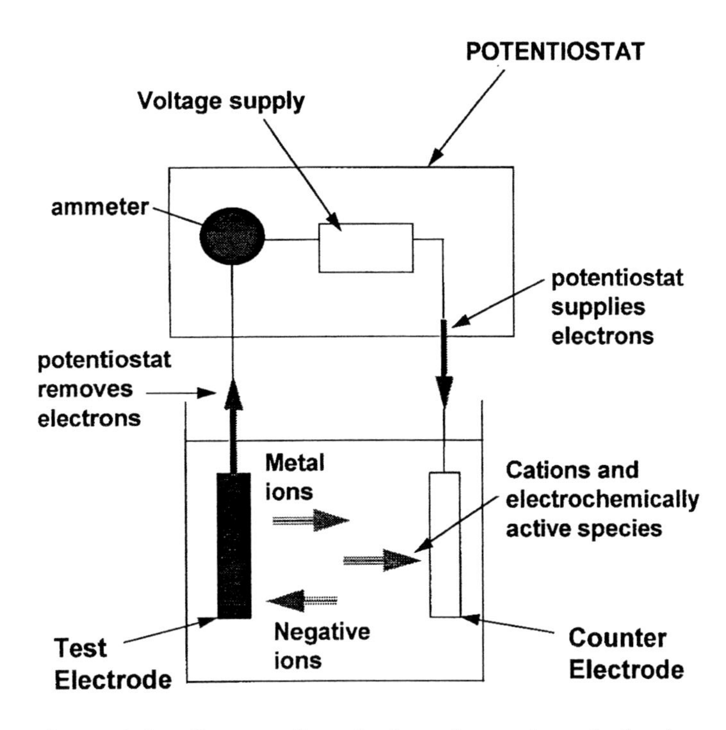

Figure 2.2. Current flow during electrode polarization

There are situations where corrosion data do not reflect actual corrosion, because one or more of the processes depicted in Figure 2.2 do not behave ideally. For example, if a test electrode metal is capable of losing 2000 electrons per second (imagine an exit ramp having more lanes than the highway), but the corresponding counter electrode can only supply 1000 electrons per second for reduction of electrochemically active species (imagine the corresponding entrance ramp having only one lane), then the potentiostat will only be able to remove 1000 electrons per second from the test electrode, even though twice that number of electrons per second could be removed.

The next section discusses how properties of an electrolyte can limit the rate of current removed from a test electrode by a potentiostat, thereby affecting electrochemical data and its correlation with actual corrosion behavior and rates.

2.5 How solution resistance affects electrochemical corrosion data

Imagine how a slow moving car causes all other cars in rush-hour traffic to reduce their speed, thus reducing the number of cars per hour that are able to enter and subsequently leave the highway. Slow moving ions, much like a slow moving car, can limit the amount of electrical current removed from (or supplied to) an electrode.

At open circuit potential (potentiostat current is zero) ions only need to move relatively short distances between anodic and cathodic sites on a corroding metal surface to maintain metal electrical neutrality, so ion motion has little or no affect on corrosion. But during electrode polarization, ions often travel greater distances along current paths established by the electrical field between test and counter electrodes in an electrolyte. Ion velocity along these current paths is determined by ion mobility (see page 18), which is a function of ion concentration, the magnitude of ion charge and frictional drag on the ion as it moves through the electrolyte.[6]

Ideal electrolyte ions move so rapidly along current paths that ion motion does not limit potentiostat electrical current. Real electrolyte ions, however, can have difficulty moving along current paths and part of the applied potential is used to move (and keep moving) ions between electrodes instead of polarizing the test electrode. Consequently, the applied potential is different than the polarization potential and some method must be employed to ensure that test electrode potentials are at desired levels. Thus the need for a reference electrode to monitor and provide feedback for adjusting test electrode potential.

The ideal way to measure test electrode potential during polarization, is to place the reference electrode on the test electrode surface. Yet, putting a reference electrode on a test electrode surface would block the flow of ionic current to (or from) the test electrode at the point of contact and could also induce pitting corrosion.[7] Consequently, the reference electrode must be placed some distance, typically two reference electrode diameters,[8,9] away from the test electrode surface to prevent test electrode blocking (also referred to as shadowing).

Unfortunately, separating the reference and test electrodes creates a potential difference between them, referred to as the IR–drop, that prevents measurement of actual electrode polarization. The electrical resistance associated with an IR–drop is referred to as uncompensated solution resistance and the mathematical expression for uncompensated solution resistance is:[10]

$$R_{\Omega} = (d)(R_{sol}) / A \qquad [2.1]$$

Where:

- R_Ω is the uncompensated solution resistance in ohms
- d is the distance between the test and reference electrodes in cm,
- R_{sol} is the solution resistivity in ohms•cm,
- A is the electrode area in cm^2

Uncompensated solution resistance, R_Ω, can be visualized as another resistor in the simple electrical circuit diagram discussed in Chapter 1 (Figure 1.3 on page 8). The simple circuit with an additional resistor looks like the circuit schematic in Figure 2.3.

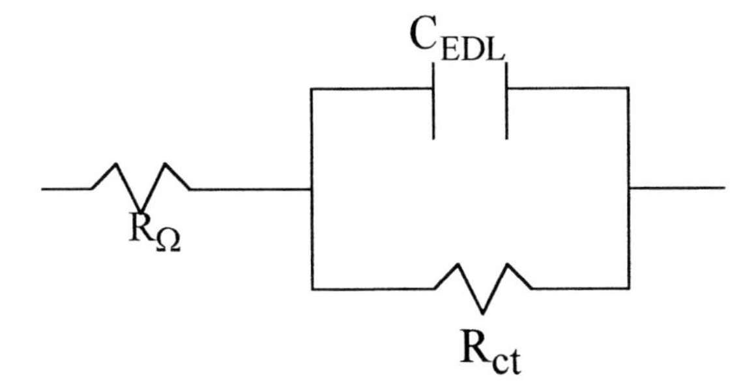

Figure 2.3. Simple electrical circuit with added uncompensated solution resistance

It can be seen by comparing Figure 2.3 with Figure 1.3 (page 8) that the total resistance for a circuit with uncompensated solution resistance (Figure 2.3) is R_Ω ohms higher than circuit resistance without uncompensated solution resistance (Figure 1.3). The mathematical expression for potentiostat voltage with uncompensated solution resistance is:

$$V_{applied} = \eta + V_\Omega \qquad [2.2]$$

Where:

- η is the test electrode overpotential (Equation 1.11 on page 11)
- V_Ω is the voltage that arises from uncompensated solution resistance, or IR–drop

The magnitude of the IR–drop (V_Ω) is equal to the electrical current supplied by the potentiostat times uncompensated resistance. Equation 2.2 demonstrates that electrode voltage at any given potentiostat current is less than the total applied potentiostat voltage when there is uncompensated solution resistance.

Corrosion data may not correspond to actual corrosion behavior or rates when the data set is composed of $V_{applied}$–current values instead of η–current values. For example, Mansfeld demonstrated that measured currents for a given applied voltage were significantly lower than actual currents when potentiostat voltages were not compensated for V_Ω.[11] Error such as that reported by Mansfeld causes underestimation of the corrosion rate and an underestimated corrosion rate can lead to a design without sufficient allowance for metal thinning by corrosion. Such an underestimation could lead to premature failure of a metal structure. Consequently, a method is needed that will allow correction of electrochemical corrosion data for uncompensated solution resistance.

2.6 Correcting electrochemical corrosion data for uncompensated solution resistance: current interrupt method

The current interrupt method capitalizes on the fact that the magnitude of V_Ω is zero when no current is supplied by the potentiostat (V_Ω equals potentiostat current multiplied times R_Ω) and provides a means of determining how much of the potentiostat potential is due to uncompensated solution resistance. The potentiostat is programmed to periodically remove (interrupt) potentiostat current for a short time (typically a few nanoseconds) and subsequently measure test electrode potential while current is zero.[12] The difference between test electrode potential with and without potentiostat current is V_Ω. Potentiostat programming either adjusts applied voltage by the amount V_Ω, thus compensating for the uncompensated solution resistance, and the potentiostat resumes test electrode polarization, or subtracts V_Ω from the data set.

There are situations where solution resistance and potentiostat currents are so low that V_Ω is negligible and the current interrupt method need not be used to correct applied potential. For example, a solution resistance of 300 ohms and potentiostat currents of 10^{-8} amps have an associated V_Ω of only $3x10^{-6}$ volts.

Using current interrupt, in this example, can produce a large amount of electronic noise in current–potential data, such as that illustrated by the top cyclic polarization curve in Figure 2.4. Whiskers in the top curve are electrical noise caused by the current interrupt procedure. Notice that noise magnitude for the top curve is less when currents are 10^{-6} amps/cm^2 or greater. The bottom curve in Figure 2.4 illustrates what the curve looks like without current interrupt. Notice how much smoother the bottom curve is when current interrupt is not used.

Electrochemical impedance spectroscopy (discussed in Chapters 7 and 8) can be used to measure uncompensated solution resistance.

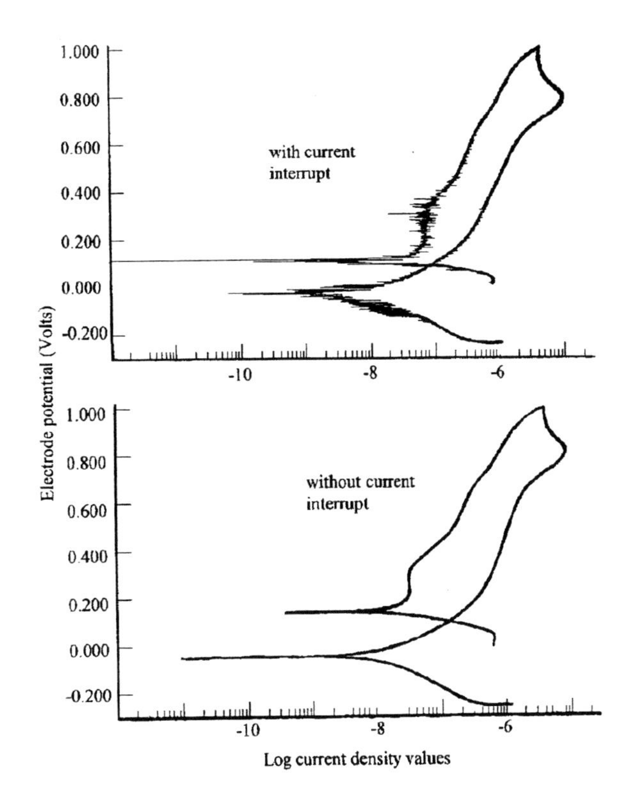

Figure 2.4. Noise produced by current interrupt when both solution resistance and electrical current are low.

2.7 Test electrode area

Equation 2.1 shows that solution resistance can be lowered by: a) increasing electrode area, b) decreasing the distance between electrodes, or c) decreasing solution resistivity.

Solution resistivity can be decreased by adding water-soluble, inorganic salts. Unfortunately, doing so can dramatically change metallic corrosion behavior. For example, adding a large amount of sodium chloride to deionized water will decrease solution resistance, but can also change stainless steel corrosion behavior from passive (no corrosion) to active (corrosion).

It has already been mentioned that there is a limit as to how close a reference electrode can be located to a test electrode surface. Thus, decreasing the reference–test electrode distance does not completely eliminate uncompensated solution resistance.

Consequently, increasing electrode area is the only remaining practical and least intrusive means for further reducing uncompensated solution resistance. Of course, test cell volume and potentiostat maximum current output will limit how much test electrode size (area) can be increased, and the experimenter must decide what is the largest test electrode size that can be used for a given test (measurement). The Tait cell on the cover of this book (EG&G Princeton Applied Research catalogue number K0307) has a 31.67 cm^2 test electrode area and has been successfully used for a variety of metals and corrosive environments (electrolytes).

2.8 Counter electrode area

Imagine how closing entrance ramp lanes can reduce the number of cars/hour entering a highway and subsequently reduce the number per hour leaving. Using a small counter electrode is analogous to closing highway entrance ramp lanes. Limiting counter electrode area, or reducing access to a counter electrode surface, reduces sites for electrochemically active species reduction by electrons and correspondingly limits potentiostat electrical current withdrawn from (or supplied to) a test electrode.

Large surface area counter electrodes, such as sintered platinum, are often used to minimize current limitation by electrode area.[13] A counter electrode whose geometrical area is at least twice that of the corresponding test electrode is also often used when it is not practical to use sintered platinum counter electrodes.

2.9 Test electrode geometrical shape

Polarized electrode electrical neutrality is maintained by movement of positive and negative ions along current paths, established by an electric field between polarized test and counter electrodes in an electrolyte. Electrochemically active species can also travel along current paths, particularly when they are ions like hydrogen ions.

Current paths are uniform and linear when polarized electrodes have simple geometry, such as that shown in Figure 2.5. Current density for this simple geometry is the total potentiostat electrical current divided by electrode area (amps per area).

Unfortunately electrode geometry is not always as simple as that in Figure 2.5, often making current distribution quite complicated,[14,15] and complicated current distributions can lead to electrochemical data that may not represent actual corrosion behavior or rates.

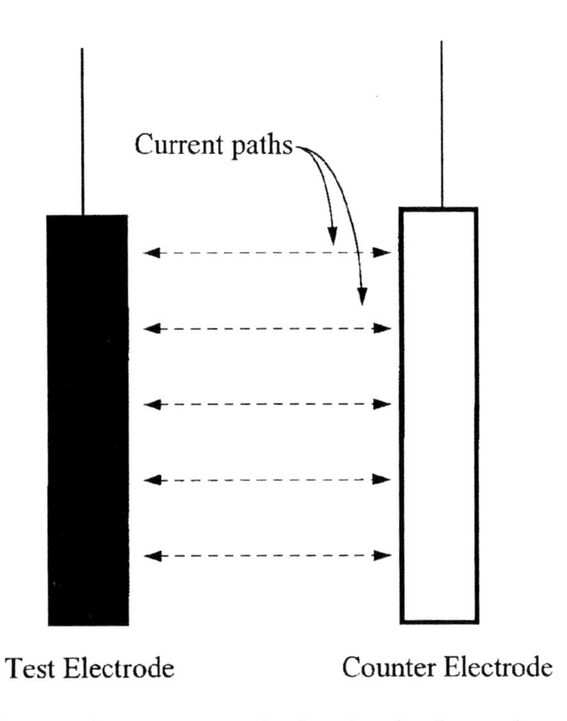

Figure 2.5. Ionic current paths for simple electrode geometry

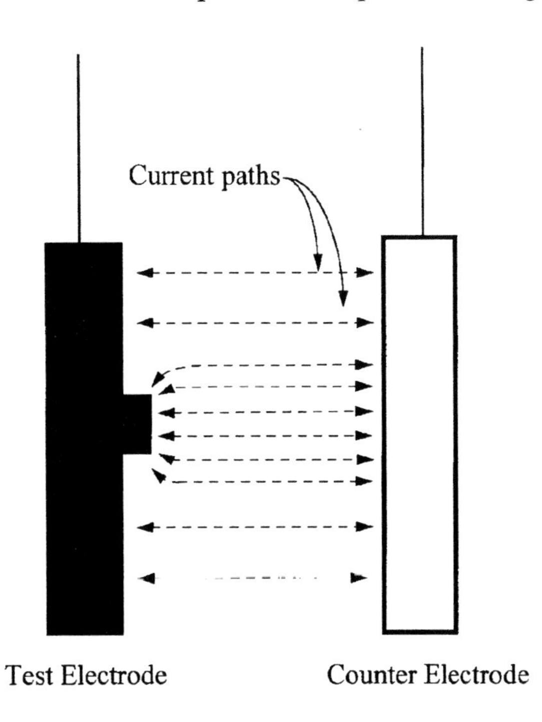

Figure 2.6. Ion current paths for complex electrode geometry

Consider the two-electrode schematic shown in Figure 2.6. One of the electrodes has a ridge on it that shortens the distance between electrodes at that location. A ridge can cause high current densities as illustrated in Figure 2.6, particularly when ions have low mobilities (i.e., low limiting velocity in the electric field).[16] High current densities around the ridge would not be expected to occur when ion mobilities are high, because ion movement between electrodes (during polarization) in this situation would be so rapid that path length between electrodes would be inconsequential.

It is important to remember that: a) higher current density does not occur at electrode surface ridges when an electrode is at its open circuit potential, and b) higher current density (during polarization) at a ridge can create localized corrosion that does not naturally occur.

Figures 2.5 and 2.6 illustrate that test and counter electrode geometry should be selected to minimize areas of abnormally high current densities on electrode surfaces. Test and counter electrodes whose opposing faces are flat and parallel are preferred because of their simple geometry. However, Annand and Eaton argued that a linear arrangement of cylindrical test and counter electrodes would also have uniform current densities and minimize occurrence of high localized currents during polarization.[16] It is also reasonable to conclude from this discussion that electrode surface finish should also be as smooth as is economically possible. It is this author's preference to use test and counter electrodes having 600 grit surface finishes.

Test cell gaskets can also form crevices at the point of contact between the gasket and electrode, and these crevices can induce metal crevice corrosion at the gasket–electrode interface.[17] It is important to use test cell gaskets that do not form crevices (between the gasket and test electrode) when there is no possibility for crevice corrosion in the corresponding real system.

2.10 Electrochemical corrosion data variability

Data variability will be observed even when all the experimental factors discussed so far have been accounted for in the test (measurement) design. Variability often makes it difficult to use electrochemical corrosion data to reach conclusions about what type of corrosion behavior or rate to expect from an actual system.

Figures 2.7 and 2.8 contain examples of electrochemical variability. Figure 2.7 contains 30 replicate anodic polarization curves from uncoated 1018 mild steel, and Figure 2.8 contains 99 replicate electrochemical impedance spectra from coated metal samples.[18]

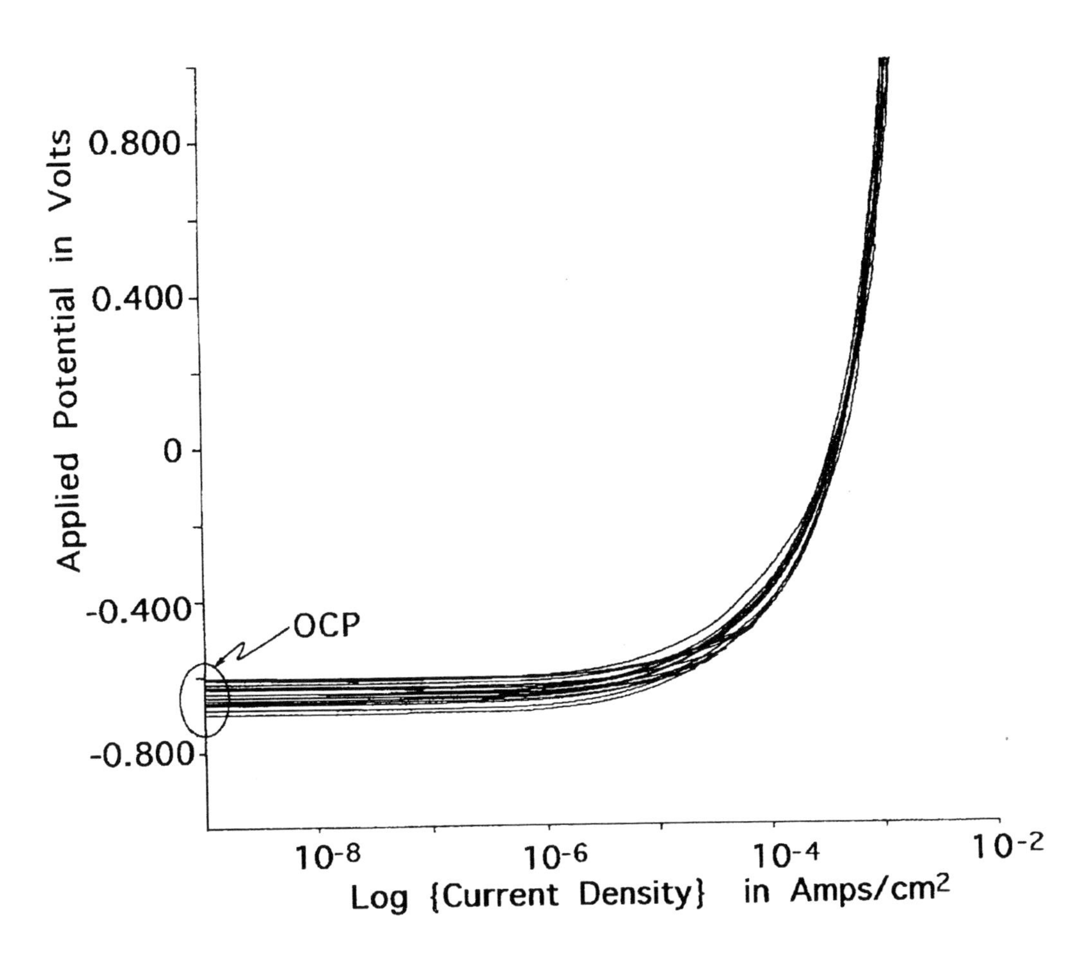

Figure 2.7. Variations in anodic polarization electrochemical corrosion data

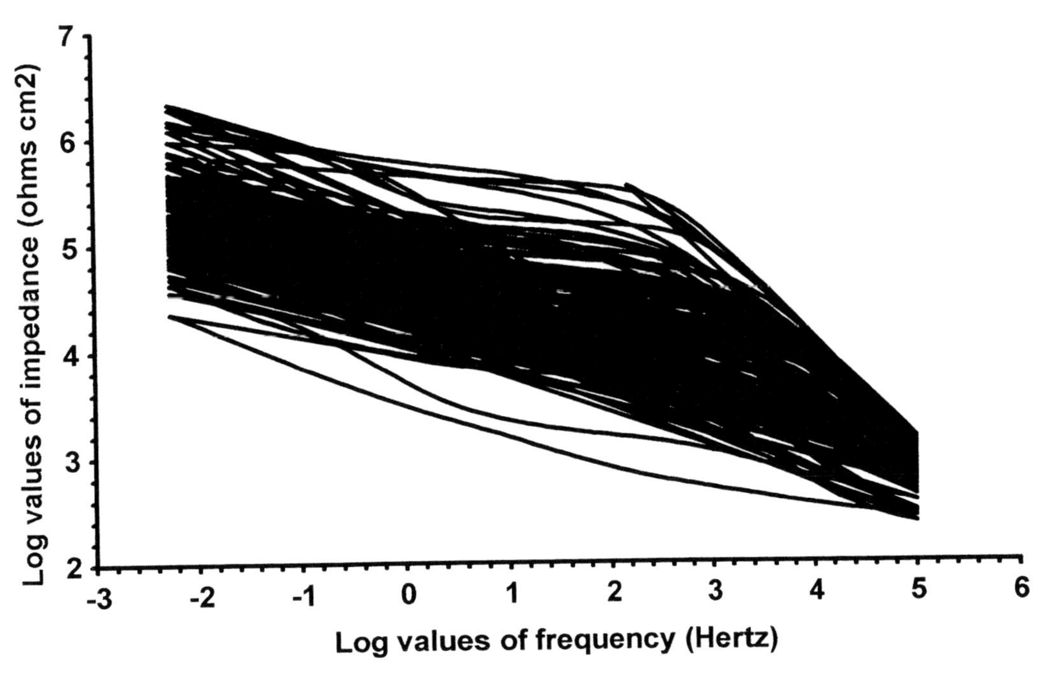

Figure 2.8. Variations in electrochemical impedance spectroscopy data

A range of current density values and open circuit potentials are observed in Figure 2.7, even though all test electrodes came from the same production batch and test electrolyte (tap water) was drawn for each test cell from the same source, at the same time. A range of impedance values for coated metals is also observed in Figure 2.8, even though all test electrodes were from the same production batch and test electrolyte was prepared from 18 megaohm deionized water and reaction grade potassium chloride.

Variations in corrosion data, such as that in Figures 2.7 and 2.8, can arise from a) measurement errors, b) chemistry variation on an electrode surface, c) localized coating chemistry variation, and d) the inherent variation of nature.

Variations in corrosion data confront the corrosion engineer or scientist with two practical questions:

1. How many repetitions should be made for each metal–environment variable?
2. How should such data be analyzed?

2.11 How many repetitions for each metal-electrolyte variable?

The question regarding the number of repetitions necessary for each metal–electrolyte variable is the most difficult to answer. The error in the mean value (for normally distributed data) for a group of repetitive measurements is estimated from Equation 2.3.[19]

$$Er^{*} = Z\sigma/\sqrt{n} \qquad [2.3]$$

Where

- Er^{*} is the difference between the true mean and the mean estimated from a group of repetitions
- σ is the standard deviation of the repetitions
- **n** is the number of repetitive measurements
- Z is the desired statistical confidence level

Equation 2.3 shows that error decreases with a) increasing sample size b) decreasing standard deviation, and c) accepting a lower statistical confidence level. Unfortunately, the only conclusion drawn from Equation 2.3 is that the number of repetitions depends on how much error one is willing to accept.

But all is not lost! Examining a graph of the error as a function of sample size gives a more satisfactory answer to the question. Figure 2.9 contains a graph of error as a function of replicate numbers (generated with Equation 2.3) for three different standard deviation values.

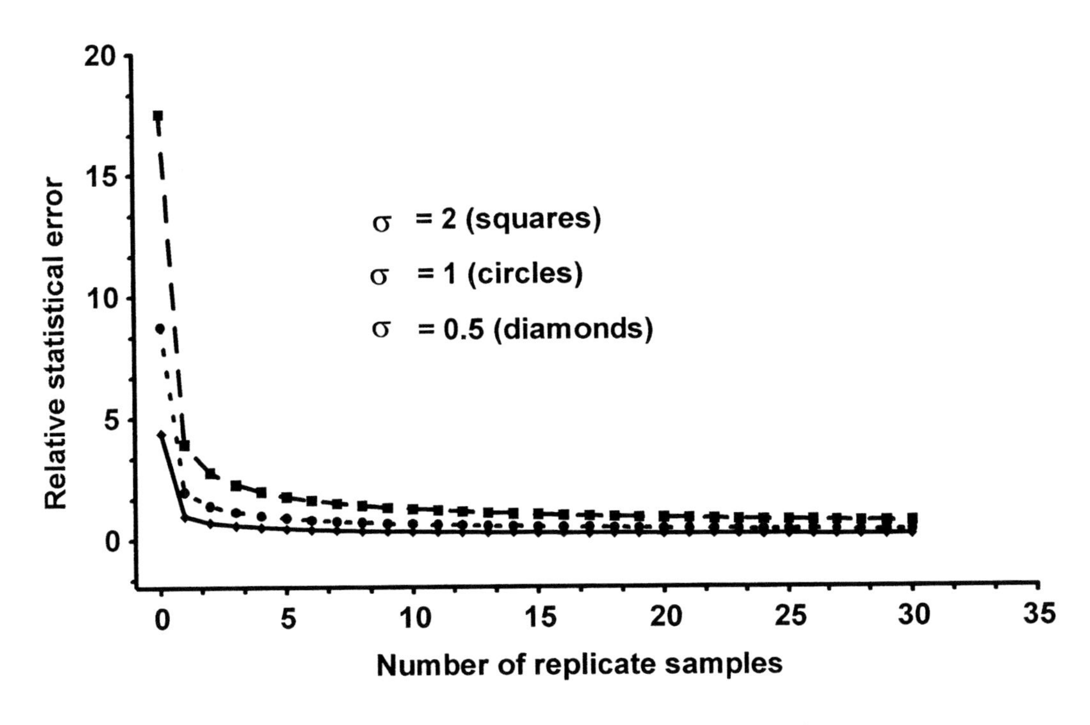

Figure 2.9. Error as a function of sample size

It can be seen in Figure 2.9 that a) the greatest rate of decrease in relative error occurs when replicate numbers are less than 5, and b) the affect of standard deviation on error magnitude becomes smaller when replicate numbers are ≥5. Consequently, a minimum of 5 replicates is recommended for electrochemical corrosion measurements.

2.12 How should electrochemical corrosion data be analyzed?

There many ways to analyze electrochemical corrosion data and the remainder of this Chapter is devoted to a brief discussion on various statistical methods that can be used for data analysis. The objective of this discussion is to introduce the Reader to statistical analysis tools that can be used to extract the most useful information from corrosion data.

A. Mean values, scatter diagrams and extreme values

It has been observed that corrosion data are not normally distributed, but are instead log normally distributed.[20] Statistical functions in hand-held calculators typically are for data that is normally distributed. Consequently, it is recommended that corrosion data be converted to a log format before using hand-held calculator statistical functions to calculate mean values. Using calculator statistical functions on unconverted data can result in erroneous mean values. Fortunately most electrochemical corrosion software reports current values in the log (base 10) format, making it unnecessary to convert corrosion currents prior to calculating mean values.

There is an easier way to analyze corrosion data other than calculating mean values. Figure 2.10 contains an example scatter diagram[18] for electrochemical impedance spectroscopy, aerosol container corrosion resistance values.

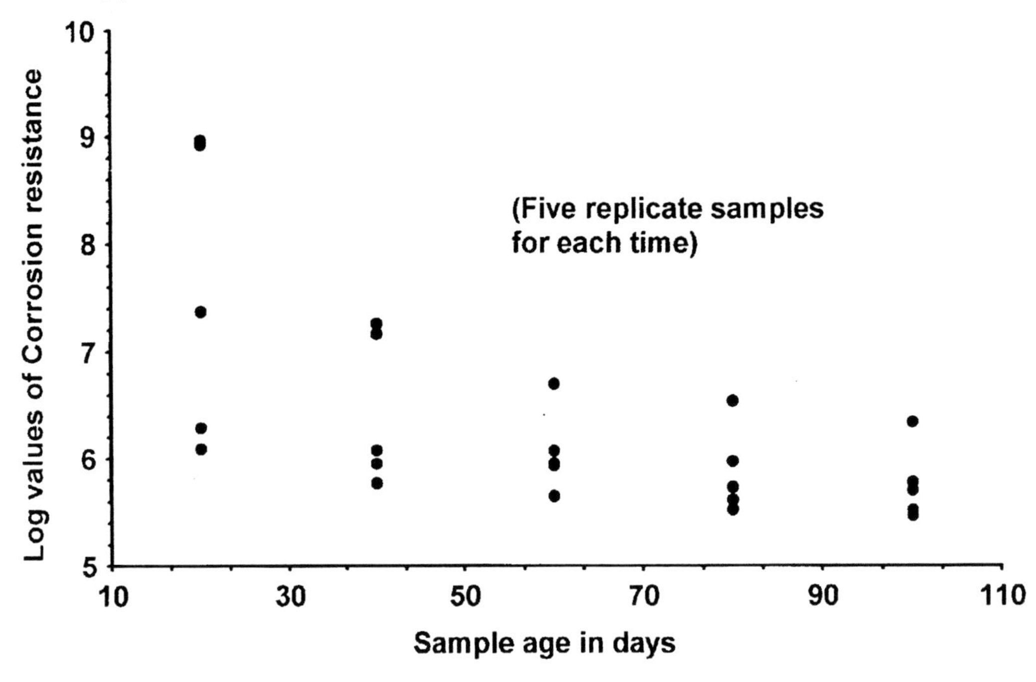

Figure 2.10. Example scatter diagram

Scatter diagrams are very powerful, simple, yet often ignored statistical analysis tools. A scatter diagram contains all repetitions for each variable, thus allowing trends to be readily seen at a glance, and estimations can be made from observed trends. For example, it was observed[21] that aerosol containers perforated within 76 days when corrosion resistance values decreased below 10^7 ohms•cm^2. It can be seen in Figure 2.10 that corrosion resistance values for all five repetitions decrease below 10^7 ohms•cm^2 within 60 days, leading to the conclusion that a significant number of containers will fail approximately 136 days (76 + 60 days) after being filled. It was reported that 100% of these containers perforated within 100 days after being filled; confirming the conclusion drawn from the data trend in Figure 2.10.[22]

Scatter diagrams allow one to analyze data without having to convert it to another format, calculate mean values, or make assumptions about what type of probability density function (statistical distribution) should be used to calculate mean values. Scatter diagrams also allow observation of extreme values.

Extreme values (often erroneously referred to as "flyers" or "outliers") contain extremely important corrosion information.[20] Extreme values, such as low corrosion resistance, or high corrosion rates, can cause early failure of metallic structures. Extreme values can only be discarded from data if they are the result of known experimental or measurement errors.

B. Nonparametric statistical tests for comparing data sets

Nonparametric statistical tests can be used to determine if results from two variables are identical without having to: a) determine what type of statistical distribution applies to the data, or b) use a large number of repetitions (e.g., 100) so that normal statistical tests (e.g., t-test) can be applied to the data. For example, the Mann-Whitney test can determine if corrosion currents for type 304 and type 316L stainless steels are statistically the same when the number of repetitive measurements for each alloy is small (e.g., five repetitions). Normal statistical tests (e.g., t-test) would require a larger number of samples for the same determination. Reference numbers 23 and 24 at the end of this chapter contain discussions on how to use nonparametric statistics, such as rank order tests (like the Mann-Whitney test) and Chi Square.

C. Regression analysis (modeling or curve fitting)

Regression analysis (often referred to as curve-fitting) is the most common type of statistical analysis used on corrosion data. Regression analysis should always be accompanied with a) analysis of variance (ANOVA), and b) analysis of residuals.[25] Regression equations (models) reported without these two statistics have not been properly validated and should be viewed with skepticism. Reference 25 contains a complete discussion on various linear and non-linear regression methods.

This brief discussion on statistics is included for two reasons:

1. to reinforce the need for corrosion scientists and engineers to make repetitive electrochemical corrosion measurements
2. to encourage engineers and scientist to use the proper statistical analyses techniques.

2.13 Summary

This Chapter discussed how uncompensated solution resistance can adversely affect electrochemical corrosion data and how experimental factors can be controlled to minimize that affect. Minimization of adverse affects on data from electrode geometry (area and shape) were also discussed. Data variability exists in most electrochemical corrosion measurements and is best addressed through use of repetitive measurements for each metal–electrolyte variable. Finally, the Reader was introduced to several statistical methods for analyzing electrochemical corrosion data.

2.14 References

1) A. J. Bard and L. R. Faulkner, Electrochemical methods: Fundamentals and Applications, p. 19, John Wiley and Sons, Inc., (1980)
2) J. O'M Bockris and A. K. N. Reddy, Modern Electrochemistry, volume 2, p. 1058, Plenum Press, NY (1970)
3) K. J. Vetter, Electrochemical Kinetics, p. 8, Academic Press, NY (1967)
4) K. J. Vetter, Op. Cit., pp. 35-73
5) A. J. Bard and L. R. Faulkner, Op. Cit., p.52
6) A. J. Bard and L. R. Faulkner, Op. Cit., p. 64
7) H. Kaesche, Metallic Corrosion, p. 311, National Association of Corrosion Engineers, Houston, TX (1985)
8) A. J. Bard and L. R. Faulkner, Op. Cit., p. 24
9) S. Barnartt, J. Electrochem. Soc., **106**, p. 724 (1959)
10) W. C. Ehrhardt, in The Measurement and Correction of Electrolyte Resistance in Electrochemical Tests, **STP 1056**, L. L. Scribner and S. R. Taylor, Eds., ASTM, Philadelphia, PA, p. 32 (1990)
11) F. Mansfeld, Electrochemical Techniques, edited by R. Baboian, p. 69, National Association of Corrosion Engineers, Houston, TX (1986)
12) N. X. Berke, D. F. Shen and K. M. Sandberg, in The Measurement and Correction of Electrolyte Resistance in Electrochemical Tests, **STP 1056**, L. L. Scribner and S. R. Taylor, Eds., ASTM, Philadelphia, PA, p. 192 (1990)
13) ASTM G 5, ASTM 1988 Annual Book of ASTM Standards, **3**, p.99, American Society for Testing and Materials, Piladelphia PA (1988)
14) F. A. Lowenheim, Electroplating: Fundamentals of Surface Finishing, p. 15, McGraw-Hill, NY (1978)
15) J. A. Faunhofer, Basic Metal Finishing, p. 19, Chemical Publishing Company, NY (1976)
16) R. R. Annand and P. E. Eaton, paper number 4, Corrosion/73, National Association of Corrosion Engineers, Houston, TX (1973)
17) J. R. Meyers, E. G. Gruenler and L. A. Smulczenski, Corrosion, **24** (10), pp. 352-353 (1968)
18) W. S. Tait, J. Coatings Technol., **66** (834), p. 59 (1994)

19) I. Miller and J. E. Freund, Probability and Statistics for Engineers, third edition, p. 188, Prentice-Hall, Englewood Cliffs, NJ (1985)

20) W. S. Tait, K. A. Handrich, S. W. Tait, and J. W. Martin, **STP 1188**, p. 428, J. R. Scully and D. Silverman, Eds., American Society for Testing and Materials, Philadelphia, PA (1993)

21) W. S. Tait, J. Coat. Technol., **62** (781), pp. 41-44 (1990)

22) W. S. Tait and K. A. Handrich, Corrosion, **50** (5), pp.373-377 (1994)

23) I. Miller and J. E. Freund, Op. Cit., pp. 271-288.

24) S. Siegel and N. J. Castellan Jr., Nonparametric Statistics for the Behavioral Sciences, second edition, McGraw-Hill Book Company, NY (1988)

25) N. Draper and H. Smith, Applied Regression Analysis, second edition, John Wiley & Sons, NY (1981)

CHAPTER 3

Time Trends in Corrosion

Objectives

After completing this Chapter, you will understand:

- that corrosion potentials and rates can change with time
- how time trends (changes) in corrosion rates affect predictions on long-term service lifetime
- how to determine when to make corrosion measurements

3.1 Introduction

Electrochemical corrosion measurements can be made any time after a metal has been submerged in an electrolyte. However, initial corrosion rates are typically different from rates measured at later times. It may not be necessary to consider corrosion time trends when one is trying to predict short-term (e.g., one week) corrosion behavior, but corrosion data from a test electrode that is not equilibrated with its electrolyte can lead to erroneous predictions of long-term service lifetime for metal structures. This chapter discusses how corrosion changes with time and how to determine the length of time a test electrode should equilibrate before making an electrochemical corrosion measurement.

3.2 More Corrosion Terminology

New terms that will be used in this chapter are defined below, and the Reader is also encouraged to review definitions presented in Chapters 1 and 2.

<u>Mils per year (mpy)</u>

Mils per year is the penetration rate of corrosion through a metal. One mil is 0.001 inches.

There are other corrosion rate terms, such as milligrams per square decimeter per day, but mils per year is the most common term found in corrosion journals.

<u>Steady state</u>

Steady state occurs when test electrode corrosion does not significantly change with time. For example, corrosion is at steady state when the corrosion rate no longer significantly changes as time increases.

Service Lifetime

Service lifetime is the length of time prior to failure that a metallic system is exposed to an electrolyte.

Service lifetime is determined by a number of factors such as 1) corrosion rate, 2) metal thickness and 3) what the user considers to be system failure. For example, a six mils per year (mpy) pitting rate may be negligible for a vessel having 1/4 inch (250 mils) wall thickness, but the same corrosion rate would be significant for a metal container having 8 mils wall thickness. Also, pin-hole leakage may or may not be considered a failure of the vessel, but would be a metal container failure. The vessel in this example would have approximately forty two years service lifetime (250 mils/6 mpy = 41.7 years) if pin-hole leakage is considered failure, but the container service lifetime would be slightly over one year (8 mils/6 mpy = 1.3 years).

Service lifetime can also be the time it takes for corrosion to reduce metal thickness to a point where a metal structure no longer has the mechanical strength needed to continue the service for which it is designed. For example, a bridge support beam may be considered to have failed when it will no longer support its designed weight (load), and its service lifetime would be the number years it took corrosion to reduce the thickness to the point where the beam can longer support its designed load.

One of the easiest corrosion parameters to measure is the open circuit potential (OCP) of a test electrode, and OCP can be used to determine when a test electrode has reached steady state.

3.3 Open circuit potential(OCP) behavior with time

Recall the Nernst equation for the relationship between electrical double layer (EDL) chemical composition and measured or applied potential.

$$E = E^{o} - (RT/nF)\ln\{\gamma_{+n}[Me^{+n}]\} \qquad [3.1]$$

The Nernst equation illustrates that magnitude of potential, E, is determined by EDL metal ion concentration and metal ion activity coefficient, γ_{+n}. Thus, any changes in either (or both) of these parameters will cause a change in the magnitude of measured potential.

OCP values change when a metal is initially exposure to an electrolyte, because it takes a finite amount of time to transform an air-formed oxide film on a metal surface into an electrical double layer (EDL). The direction of potential change is determined by how EDL chemistry adjusts to accommodate electrolyte chemistry. That is, does the metal surface form a passive film that protects the metal from further corrosion (typically an increasing OCP); or does the metal surface form a porous hydroxide layer that only slows the rate of corrosion, but does not protect the metal from further corrosion (typically a decreasing OCP)?

Figures 3.1 and 3.2 contain examples of OCP–time behavior for type 1018 mild steel in an emulsion and type 1100 aluminum in water, respectively. Even though steel OCP decreases as a function of time, and aluminum OCP increases as a function of time, both reach a time where no significant changes in OCP values are observed.

The time it takes to reach steady state depends in part on the type of metal. Aluminum in Figure 3.2 reaches steady state faster than steel in Figure 3.1, around two hours as opposed to approximately 16 days. Electrolyte chemistry also influences the time it takes to reach steady state. For example, OCP for type 1018 mild steel reaches steady state in approximately 2 to 4 hours when submerged in oxygen saturated water,[1] as opposed to the sixteen hours observed in Figure 3.1. Other researchers have also observed changes in OCP values with time for a variety of metals and electrolytes.[2,3,4]

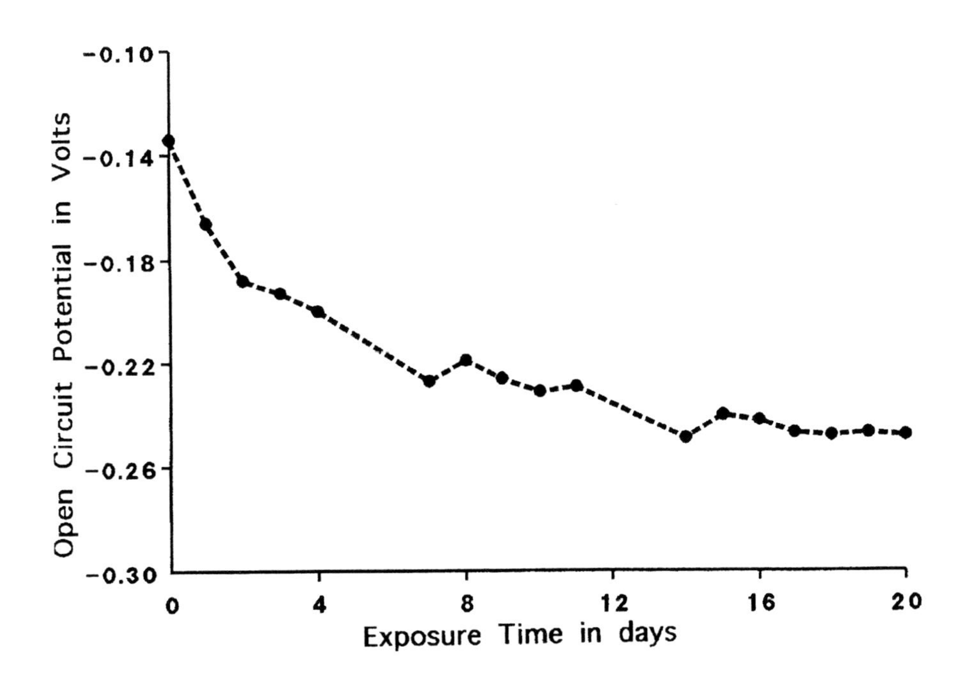

Figure 3.1. OCP-time behavior of steel in an emulsion

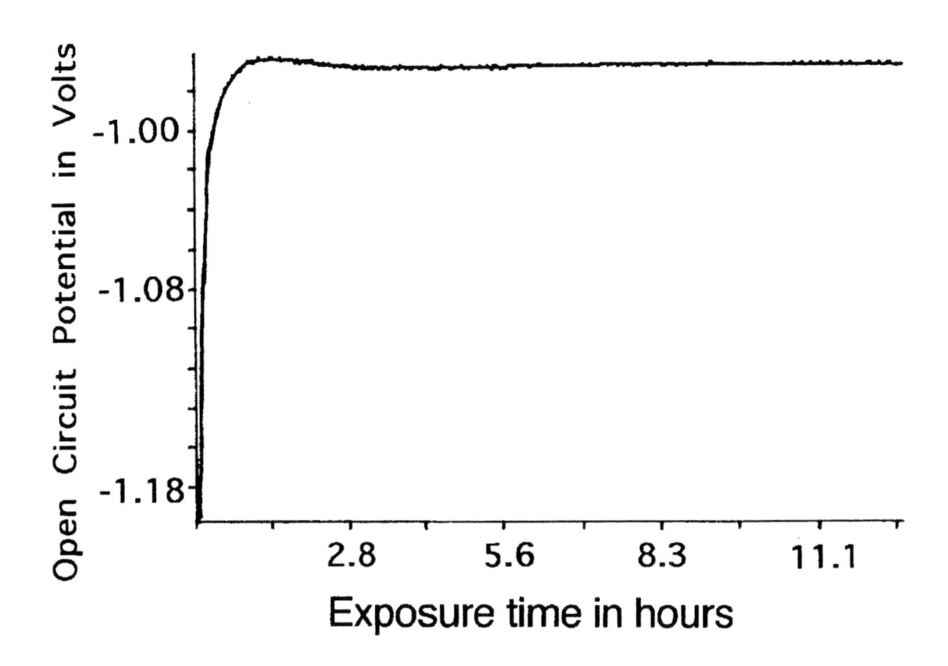

Figure 3.2. OCP-time behavior of aluminum in deionized water

Electrolyte chemical composition also influences the magnitude of metal ion activity coefficients.[5] Changing bulk electrolyte chemistry can cause the chemical composition of an electrical double layer (EDL) to adjust, resulting in a different OCP. For example, decreasing electrolyte pH from 9 to 2 (e.g., by adding hydrochloric acid) changes EDL chemistry by increasing hydrogen ion concentration in the EDL and thus changes the corresponding OCP. Consequently, it typically requires time for the electrode to reach steady state again when bulk electrolyte chemistry changes.

Temperature will also affect the magnitude of metal ion activity coefficients in Equation 3.1 (γ_{+n}), and thus time must be allowed for a test electrode to reach steady state when system temperature is changed. However, it has been this author's experience that typically only short times are needed to reach steady state after temperature changes.

The relationship between potential and current was discussed in Chapter 1 (see pages 10 through 12). This relationship makes it reasonable to hypothesize that corrosion rates may also require a finite time to reach steady state magnitudes.

3.4 Corrosion rate behavior with time

Corrosion rates, like OCP, typically do indeed require a finite amount of time to reach steady state magnitudes. Figure 3.3 contains a graph of corrosion rate-time behavior for data tabulated by Mottern and Myers.[6] The initial corrosion rate for the data in Figure 3.3 is approximately 6 mils per year and the final rate is approximately 0.6 mils per year. Other researchers also observed that corrosion rates require a finite amount of time to reach steady state values, and that initial corrosion rates are typically higher than steady state rates, such as illustrated in Figure 3.3.[7,8,9] Mild steel pitting corrosion rates in aerated tap water were observed to require approximately four hours to reach steady state values.[4]

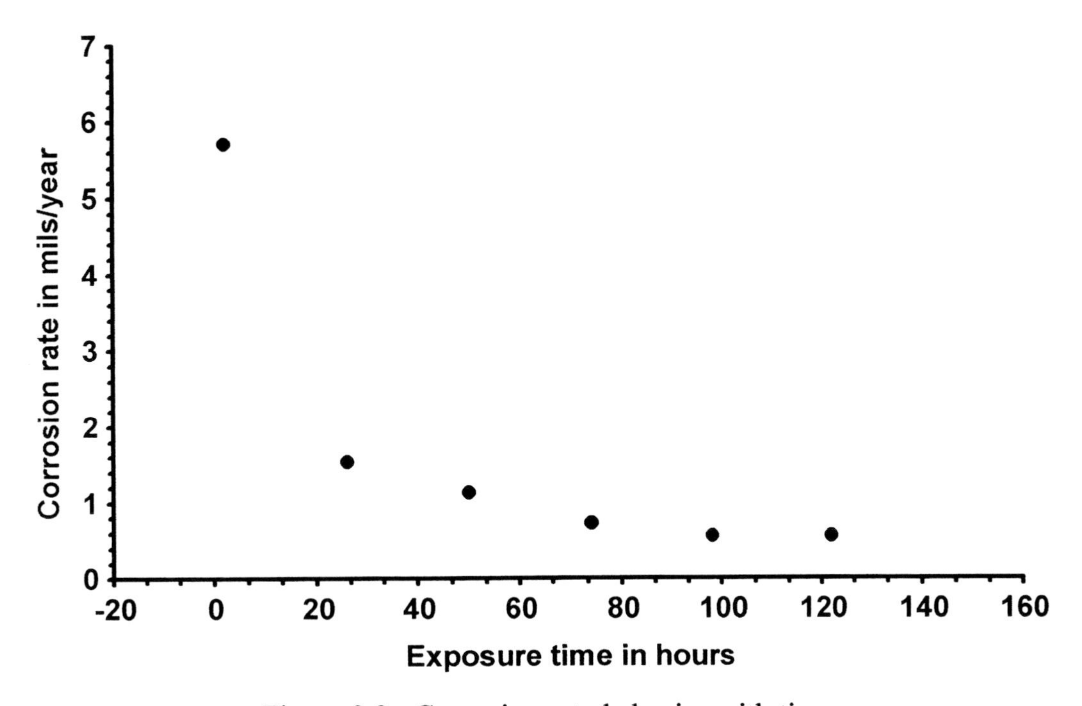

Figure 3.3. Corrosion rate behavior with time

3.5 How does time behavior impact long-term corrosion predictions?

Metal structure failure depends upon the type of service for which the structure is designed. Consequently, each corrosion engineer or scientist must determine what constitutes failure for their system, and thus service lifetime.

The service lifetime of a 100 mil thick metal structure would be approximately 17 years if a) the initial corrosion rate in Figure 3.3 (6 mpy) is used to estimate service lifetime and b) perforation by corrosion is considered failure. Conversely, the estimated service lifetime for the same thickness is 170 years when the steady state rate (0.6 mpy) is used for the lifetime calculation. Thus the initial rate gives a significantly shorter estimated service lifetime than does the steady state corrosion rate.

Metal exposure time to an electrolyte should be verified for cases of alleged short-term exposure. For example, a tank truck may be emptied after holding an electrolyte for a few days, but residual electrolyte may remain in the tank truck after it is emptied. Consequently, there would be areas of metal that are continuously exposed to electrolyte for a longer time than a few days. One may want to treat this situation as if it were a long-term exposure and use steady state data for predicting service lifetime.

3.6 Summary

Electrochemical corrosion measurements can be made at any point after a test electrode is submerged in an electrolyte. However, it may be necessary to measure corrosion at steady state when corrosion data are used to estimate long-term service lifetime. Service lifetime is different for each situation because what constitutes metal structure failure is determined by the type of service and by:

1. metallic corrosion rate
2. metal thickness
3. time that a metal is exposed to an electrolyte
4. mechanical strength needed for the desired type of service

Open circuit potential can be monitored as a function of time to determine when a test electrode is at steady state. Other corrosion parameters (e.g., corrosion rate) can also be used to determine when steady state is achieved, but OCP measurements are the simplest non-destructive measurements that can be made.

Short-term service may not necessarily be short-term exposure if residual electrolyte remains in a structure after the majority of the electrolyte has been removed. The corrosion engineer or scientist may want to use steady state corrosion rates to predict service lifetime for this type of situation.

3.7 References

1) W. S. Tait, Corrosion, **35** (7), pp. 296-300 (1979)
2) D. D. Macdonald, H. Song, K. Makela., K. Yoshida, Corrosion, **49**(1), pp. 8-16 (1993)
3) I. Sekine, et. al., Corrosion Science, **32** (8), pp. 815-825 (1991)
4) M. Janik-Czachor, Corrosion, **49** (9), pp. 763-768 (1993)
5) W. S. Tait and K. A. Handrich, Corrosion, **50** (5), pp. 373-377 (1994)
6) M. M. Mottern and J. R. Myers, Corrosion, **24** (7), pp. 197-205 (1968)
7) W. J. Lorenz and F. Mansfeld, Corrosion Science, **21** (9), pp. 647-672 (1981)
8) B. Yang, D. A. Johnson, S. H. Shim, Corrosion **49** (6), pp. 499-513 (1993)
9) H. Vedage, T. A. Ramanarayanan, J. D. Mumford and S. N. Smith, Corrosion, **49** (2), pp. 114-121 (1993)

CHAPTER 4

Linear Polarization Corrosion Measurement

Objectives

After completing this Chapter, you will know:

- how to make linear polarization measurements
- how to convert corrosion resistance values to corrosion rates
- the advantages and limitations of linear polarization measurements

4.1 Introduction

No external electrical current flows to or from a test electrode when it is at open circuit potential (OCP), even when the electrode is corroding and thus has a corrosion current, i_{corr}. An external potential must be applied to a test electrode to move its potential away from OCP and thus obtaining an electrical current that can be measured. How can applied potentials, and their corresponding currents, be used to determine the magnitude of an actual corrosion current which occurs in the absence of an applied potential? Remember the Bulter Volmer equation from Chapter 1 which relates applied potential (overpotential) to its corresponding current:

$$i = i_{corr} \{\exp(\alpha nF\eta/RT) - \exp(-(1-\alpha)nF\eta/RT)\} \qquad [4.1]$$

This equation mathematically illustrates the relationship between applied potentials (overpotentials, η), their corresponding electrical currents and corrosion current. Consequently, a spectrum of potentials and their corresponding currents can be used to determine the corrosion current density, i_{corr}, and thus the corrosion rate.

This chapter begins a three-chapter-series discussion of direct current (DC) electrochemical polarization methods.

4.2 More corrosion terminology

You are encouraged to review definitions in Chapter 1 (pages 2 and 3), Chapter 2 (pages 18 and 19), Chapter 3 (pages 37 and 38) and may also want to re-read section 2.4 of Chapter 2 (pages 20 and 21) which discusses what happens to electrodes and electrolytes during test electrode polarization.

Several new corrosion terms are defined in this section and others are presented throughout this Chapter with their associated mathematical equations.

Anodic current:

Anodic current refers to the electrical current withdrawn from a test electrode during anodic (oxidation) polarization.

Cathodic current:

Cathodic current refers to the electrical current supplied to a test electrode during cathodic (reduction) polarization.

Corrosion Resistance:

Corrosion resistance can be defined as a metal's ability to resist corrosion, or the resistance of a metal to transfer its electrons to electrochemically active species in solution. Corrosion resistance is also called polarization resistance or charge transfer resistance and has units of ohms•cm^2 or ohms.

4.3 A brief overview of DC corrosion test methods

The cyclic polarization curve in Figure 4.1 will be used for a brief overview on the DC polarization corrosion measurement methods discussed in this book, before we begin discussion on linear polarization. Applied potentials in Figure 4.1 are plotted on the Y-axis against log values of their corresponding electrical current densities on the X-axis.

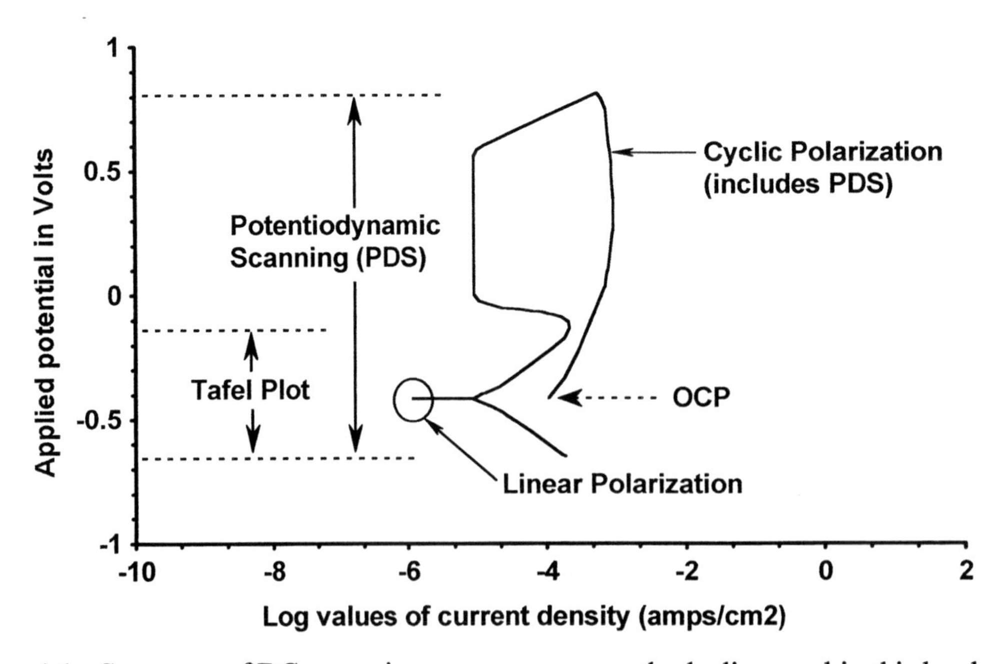

Figure 4.1. Summary of DC corrosion measurement methods discussed in this book

The linear polarization method (discussed in this chapter) uses the smallest potential spectrum of all the DC corrosion measurement methods. The circled area in Figure 4.1 depicts the location of the linear polarization potential spectrum in the cyclic polarization curve. Linear polarization measurements begin at approximately -20 mV from OCP and end around +20 mV from OCP. Unlike other DC methods, linear polarization data are plotted on a linear graphical scale and the data typically fall on a straight line. A more detailed discussion on linear polarization will follow in the next section of this Chapter.

The potential spectrum for the Tafel plot method (discussed in Chapter 5) is labeled in Figure 4.1 with double arrows next to the potential axis. Tafel plot spectra begin at approximately -250 mV from OCP and end at approximately +250 mV from OCP. Tafel plot data are plotted as potential versus log values of current density as shown in Figure 4.1. Remember, from the discussion in Chapter 1 on the Tafel plot in Figure 1.5, that a Tafel plot has anodic and cathodic branches corresponding to anodic (oxidation) and cathodic (reduction) corrosion half reactions for a metal.

The potentiodynamic scanning method (discussed in Chapter 6) uses a potential spectrum that begins around -250 mV from OCP and ends at approximately +1000 mV from OCP. The anodic branch of the potentiodynamic scan (potentials larger than OCP) is typically non-linear and in certain situations can have the S-shape depicted in Figure 4.1.

The cyclic polarization method (also discussed in Chapter 6) incorporates the potentiodynamic scan potential spectrum, plus potentials from a reverse scan that is initiated from the end of the potentiodynamic scan back to OCP, as shown in Figure 4.1.

Each method has advantages and disadvantages and one method may be more suitable for a given application than the others, or a combination of several methods may be needed for certain situations.

The remainder of this Chapter will focus on discussion of the linear polarization method.

4.4 Linear Polarization

The main advantage linear polarization has over other DC corrosion measurement methods is its potential spectrum is so small, that it is essentially a non-destructive test. Consequently, linear polarization measurements can be repeatedly made on the same test electrode, allowing it to be used for applications like a) long-term corrosion monitoring and b) determining when a test electrode is at its steady state corrosion rate.

Figure 4.2 contains linear polarization data collected with an approximately 20 mV wide potential spectrum. Test electrode electrical current is zero at OCP, as shown in Figure 4.2, and electrode potential polarity switches from cathodic to anodic as the scan proceeds past OCP.

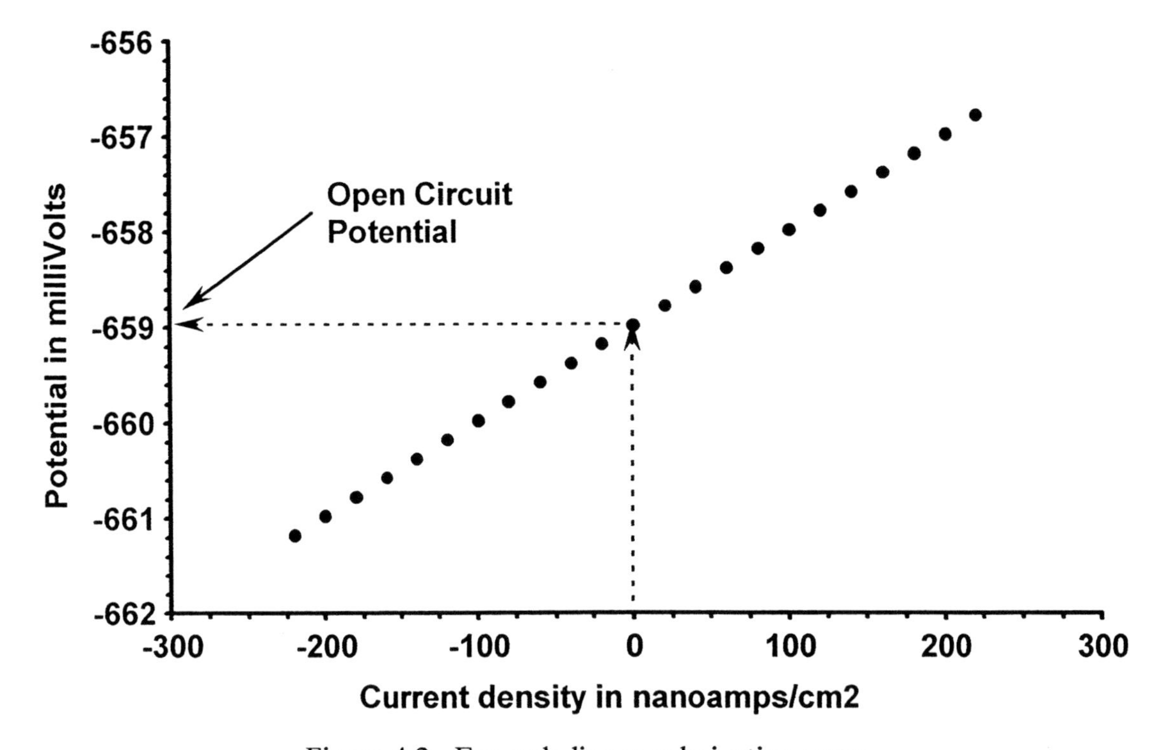

Figure 4.2. Example linear polarization curve

A potentiostat is programmed to collect linear polarization data as follows:

1. test electrode OCP is measured prior to polarization
2. the potentiostat begins test electrode polarization at approximately -20 mV from OCP and increases potential in small steps (typically 1 to 2 mV per step), until the final electrode potential is approximately +20 mV from OCP
3. electrical currents are recorded for each potential step

Linear polarization measurements are susceptible to errors caused by uncompensated solution resistance and the current-interrupt method should be incorporated into the measurement when solution resistance is high (see Chapter 2 pages 24 and 25).

The next section discusses how: a) to use linear polarization data to obtain corrosion resistance values and b) corrosion resistances are converted to corrosion rates.

4.5 Calculating corrosion currents and rates from linear polarization data

The slope of a line is the change in its Y-values divided by the change in its X-values. Hence the slope for a linear polarization curve is the change in potential divided by the corresponding change in current density. This relationship is written mathematically as:

$$\text{Slope} = \Delta E/\Delta i \qquad [4.1]$$

The slope in Equation 4.1 has resistance units in ohms•cm^2 and is referred to as corrosion, or polarization resistance, R_p.

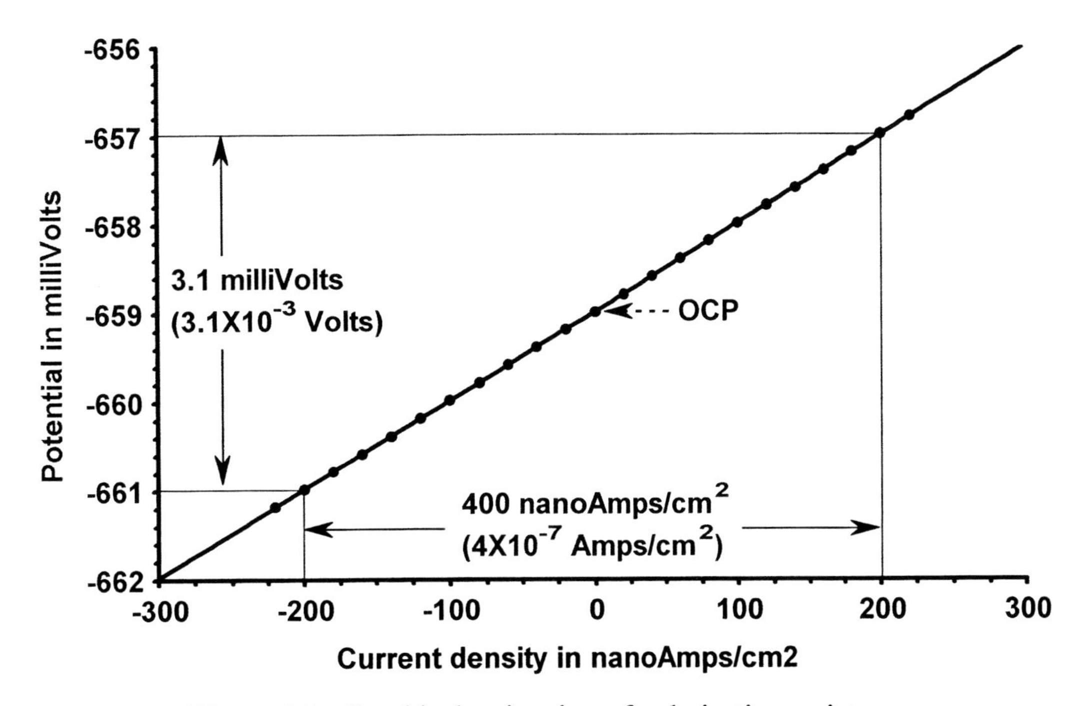

Figure 4.3. Graphical estimation of polarization resistance

Figure 4.3 contains a line fitted to the Figure 4.2 linear polarization data. The potential change for the fitted line is 3.1 mV and the corresponding current density change is 4X10^{-7} Amps/cm^2, so the corrosion resistance, R_p, is:

R_p = $\Delta E/\Delta i$

= (3.1X10^{-3} Volts)/(400X10^{-9} Amps/cm^2)

= 7,750 Volts/Amps/cm^2, but Volts/amps is ohms, so R_p in this example is,

= 7,750 ohms•cm^2

Corrosion current must be converted to a corrosion rate in order to use it for practical determinations like: a) how fast is the metal corroding, or b) estimating metal structure service lifetime. Stern and Geary developed the following mathematical relationship between corrosion resistance and corrosion current for linear polarization data.[1,2]

$$i_{corr} = [1/(2.303R_p)][(\beta_a \cdot \beta_c)/(\beta_a + \beta_c)] \qquad [4.2]$$

Where:

- i_{corr} is the corrosion current density in Amps/cm^2
- R_p is the corrosion resistance in ohms•cm^2
- β_a is the anodic Tafel slope in Volts/decade or mV/decade (1000 mV = 1 Volt) of current density
- β_c is the cathodic Tafel slope in Volts/decade or mV/decade of current density
- The quantity, $(\beta_a \cdot \beta_c)/(\beta_a + \beta_c)$, is referred to as the Tafel constant

Pages 11 through 15 of Chapter 1 discuss how Tafel plot anodic and cathodic branches in Figure 1.5 (page 13) both have a linear portion that occurs at approximately 50 mV from OCP. The slopes of these linear portions are referred to as Tafel slopes. Tafel slope values have been reported from 1 mV/decade to 105 mV/decade.[3,4] Chapter 5 contains a more complete discussion on Tafel plots and how to determine Tafel slopes.

The corrosion resistance from Figure 4.3 will be used to illustrate corrosion resistance conversion to corrosion current with Equation 4.2. We will assume that β_a = 60 mV/decade and β_c = 105 mV/decade for this calculation. Substituting values for β_a, β_c and R_p into Equation 4.2 gives:

$$\begin{aligned} i_{corr} &= [(7{,}750)(2.303)\ \text{ohm}\cdot\text{cm}^2]^{-1}[(60)(105)/(60 + 105)]\ \text{mV} \\ &= [1/17848.25]\cdot[(6300)/(165)]\ \text{mV/ohm}\cdot\text{cm}^2 \\ &= 0.00214\ \text{mV/ohm}\cdot\text{cm}^2 \\ &= 0.00214\ \text{mA/cm}^2 \end{aligned}$$

Corrosion current can be converted to a corrosion rate, in mils/year (mpy), using the following equation:

$$\text{MPY} = i_{corr}(\Lambda)(1/\rho)(\varepsilon) \qquad [4.3]$$

Where:

- Λ is a combination of several conversion terms and is $1.2866\text{X}10^5$[equivalents•sec•mils]/[Coulombs•cm•years]
- i_{corr} is the corrosion current density in Amps/cm^2 (1 Amp = 1 Coulomb/sec)
- ρ is metal density in grams/cc
- ε is equivalent weight (equivalent) in grams/equivalent. Equivalent weight is a metal's gram molecular weight divided by the number of electrons in a metal's anodic half reaction (e.g., Equation 1.1 on page 4)

If the corrosion current in our example calculation is for iron, then the density is 7.86 grams/cm^3 and equivalent weight is 27.56 grams/equivalent. Substituting values for i_{corr}, Λ, ρ and ε into Equation 4.3 gives:

$$\begin{aligned} \text{MPY} &= (2.14\text{X}10^{-6}\ \text{Amps/cm}^2)(1.2866\text{X}10^5)(27.56)(1/7.86) \\ &= 0.97\ \text{mpy} \end{aligned}$$

Page 38 of Chapter 3 contains a discussion on how to estimate service lifetime with corrosion rates.

Stern and Geary made the following assumptions when deriving their equation:

1) Corrosion reactions for the system being measured are reversible
2) Both the anodic and cathodic corrosion reactions are controlled by the corrosion reaction activation energy
3) Electrode surface changes do not occur during polarization
4) Polarization is due to corrosion
5) The energy barrier for the forward and reverse corrosion half reactions are symmetrical

These assumptions should be kept in mind when using Equation 4.2 to calculate corrosion current from linear polarization data, because many of these assumptions do not apply to practical situations. Hence, corrosion currents and rates calculated from equations 4.2 and 4.3 should be considered only as estimates, when the Stern Geary assumptions do not apply to the real system.

4.6 Practical experiences using linear polarization

Annand reported good correlation between linear polarization corrosion rates and rates obtained from weight loss data, when corrected Tafel slopes were used with the Stern Geary equation[5] (Weight loss corrosion rates are calculated by dividing the weight of metal lost from corrosion by metal exposure time to an electrolyte.). However, Callow et. al. concluded that the correlation between linear polarization and weight loss corrosion rates was tenuous, based on their work and review of linear polarization studies published in various corrosion journals.[4]

Poor correlation between linear polarization corrosion rates and rates obtained from other methods (e.g., weight loss) can originate from using wrong Tafel slope values. Table 4.1 illustrates how Tafel slopes can affect the magnitude of corrosion currents, by comparing i_{corr} values calculated with different Tafel slopes. For example, using the same resistance, but increasing the cathodic Tafel slope from 60 to 120 increases i_{corr} approximately 34% and doubling both Tafel slopes from 60 to 120 increases i_{corr} approximately 300% for the same resistance.

Table 4.1. How Tafel slopes affect the magnitude of corrosion currents

R_p ohms·cm^2	β_a mV/decade	β_c mV/decade	i_{corr} mA/cm^2
1214	60	60	10.7
1214	60	120	14.3
1214	120	120	42.9

Errors in linear polarization data can also occur when data are:

1. taken from a test electrode that is not at steady state (see Chapter 3 pages 40-42)
2. not corrected for uncompensated solution resistance (see Chapter 2 pages 22-24)
3. the scan rate used to collect the linear polarization data is faster than the rate at which test electrode electrical double layer can adjust to each potential change

4.7 Limitations

Linear polarization can only measure general corrosion rates and thus can not be used to determine: a) if localized corrosion such as pitting or crevice corrosion is present, or b) what type of kinetics are controlling the rate of corrosion. It is often difficult to obtain linear data in high resistance solutions, or for metals that have extremely low corrosion rates, even when the IR-compensation method is included in the measurement.

4.8 Summary

The main advantage of the linear polarization method is that it is a non-destructive test and can be repeatedly used on the same test electrode, to continuously monitor corrosion for long times. The other DC methods damage test electrode surface to the extent that measurements can be either only repeated a limited number of times on a test electrode because polarization causes test electrode surface roughening (Tafel plots); or only one measurement can be made on a test electrode because polarization induces pitting (cyclic polarization and potentiodynamic scanning). However, the other methods provide information that linear polarization does not provide about localized corrosion and corrosion mechanisms.

Linear polarization can be used to determine when a test electrode is at its steady state rate so that another DC measurement (e.g., cyclic polarization) can be made at steady state.

4.9 References

1) M. Stern and A. L. Geary, J. Electrochem. Soc., **104** (1), pp. 56 - 63 (1957)
2) M. Stern, Corrosion, **14** (10), pp. 61 - 64 (1958)
3) F. Mansfeld, Electrochemical Methods, edited by R. Baboian, pp. 67 - 71, National Association of Corrosion Engineers, Houston, TX (1986)
4) L. M. Callow, J. A. Richardson and J. L. Dawson, Br. Corr. J., **11** (3), pp. 123 -139 (1976)
5) R. R. Annand, 21st Annual Conference of the NACE, St. Louis, MO (1965)

CHAPTER 5

Tafel Plot Corrosion measurement

Objectives

After completing this Chapter, you will understand:

- how to generate a Tafel plot
- how Tafel plots are used to determine what processes control corrosion rates
- how to estimate Tafel slopes
- how to estimate corrosion rates from Tafel plots
- limitations on the use of Tafel plots

5.1 Introduction

Linear polarization curves (discussed in the previous chapter) are useful for situations where long-term corrosion monitoring is needed, without having to routinely replace test electrodes. However, linear polarization corrosion rates are only estimates of actual rates when Tafel slopes are unknown and it is sometimes desirable to know more about corrosion than just its rate. The Tafel plot corrosion measurement method uses a wider DC potential spectrum (400 to 500 mV) and provides more corrosion information than linear polarization.

5.2 Still more corrosion terminology

You may want to re-read Chapter 1 section 1.7 (pages 12 through 15) which discusses Tafel plots in relation to the Butler Volmer equation (page 11). You may also want to review definitions in Chapter 1 (pages 2 and 3), Chapter 2 (pages 18 and 19), Chapter 3 (pages 37 and 38) and Chapter 4 (page 44).

<u>Activation control</u>

Corrosion is activation controlled when the corrosion rate is determined by how fast a metal electrode can transfer its electrons to electrolyte electrochemically active species (EAS).

<u>Diffusion control</u>

A corrosion rate is diffusion controlled when the rate is determined by electrolyte EAS diffusion rate to an electrode surface.

Scan rate

Scan rate is the rate at which a potentiostat changes a test electrode potential. Scan rate has units of millivolts per second or Volts per second.

5.3 Generating Tafel plots

Remember from the discussion on pages 12 through 15 that a Tafel plot has anodic and cathodic branches, corresponding to the anodic and cathodic half reactions for metal corrosion. Figure 1.5 on page 13 and Figure 5.1 on the next page contain examples of Tafel plots. Tafel plots are generated in one of two ways:

Method 1

Tafel plot anodic and cathodic branches are generated from the same test electrode. Polarization is begun at approximately -200 mV from OCP and increased until the potential is approximately +200 mV from OCP.

Method 2

Separate test electrodes are used in the same electrolyte to generate Tafel plot anodic and cathodic branches. The cathodic branch is generated by polarizing one of the test electrodes from OCP to approximately -200 mV from OCP and the anodic branch is generated by polarizing the other test electrode from OCP to approximately +200 mV from OCP.

Potential changes are typically 2 mV per step for both methods and potential-current data are graphed as applied potential versus log values of current density.

Remember how the Nernst equation relates potential to electrical double layer (EDL) chemical composition (pages 9 through 11). Polarization changes EDL chemical composition and there are situations where method 1 produces a Tafel plot whose OCP is not located at the anodic and cathodic inflection point (see Figure 5.1), because of EDL composition changes. Tafel plot OCP and inflection point can be different when the scan rate is such that potentials change faster than the EDL can adjust its chemical composition for each potential.

Using method 2 can create a Tafel plot with a gap between the anodic and cathodic branches at the inflection point, making estimation of the corrosion current, i_{corr}, difficult. OCP variability discussed on pages 28 through 30 and Figure 2.7 on page 29 can cause an inflection point gap between Tafel plot anodic and cathodic branches.

5.4 Tafel plot structure

Tafel plots can be classified as either activation or diffusion controlled. Tafel plots are classified as activation controlled when the corrosion rate is determined by how fast a metal is capable of transferring its electrons to electrolyte EAS. Tafel plots are classified as diffusion controlled when EAS diffusion rate determines the corrosion rate.

Figure 5.1 contains an example of a Tafel plot where corrosion reaction activation energy controls the corrosion rate. A characteristic of activation control is increasing current density magnitudes with potential increases for both branches. Notice in Figure 5.1 that both branches become linear at approximately 50 mV from OCP. Figure 1.5 in Chapter 1 (page 13) is also an activation controlled Tafel plot.

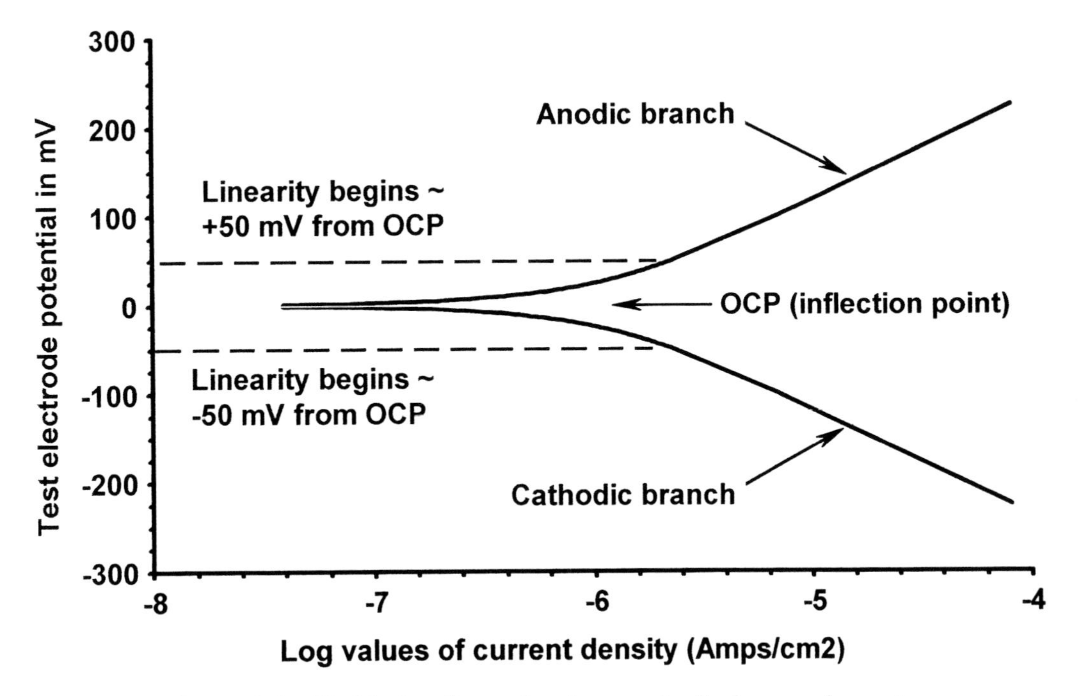

Figure 5.1. Tafel plot for activation controlled corrosion rates

Diffusion can restrict access of electrolyte EAS to an electrode surface when:

1. an electrolyte has a limited supply of EAS (e.g., at pH 14 the hydrogen ion concentration is 10^{-14} moles/liter)
2. EAS diffuse very slowly through the electrolyte to an electrode surface
3. corrosion reaction products restrict EAS access to an electrode surface

Examples of corrosion reaction products that restrict EAS surface access are: a) hydrogen gas bubbles resulting from hydrogen ion reduction (Equation 1.2 on page 4), that restrict hydrogen ion access to an electrode surface[5] and b) thick metal-hydroxide corrosion product layers that can restrict EAS diffusion to an electrode surface.[3]

Figure 5.2 contains examples of diffusion controlled Tafel plots. Diffusion control theoretically causes cathodic current density to become constant at approximately -50 mV from OCP, as illustrated by the solid (ideal) plot. However, there are diffusion controlled plots whose currents increase slightly with potential changes, as illustrated by the dashed plot. The constant, or slowly changing, cathodic current is referred to as the diffusion limited current, as noted for both plots in Figure 5.2.

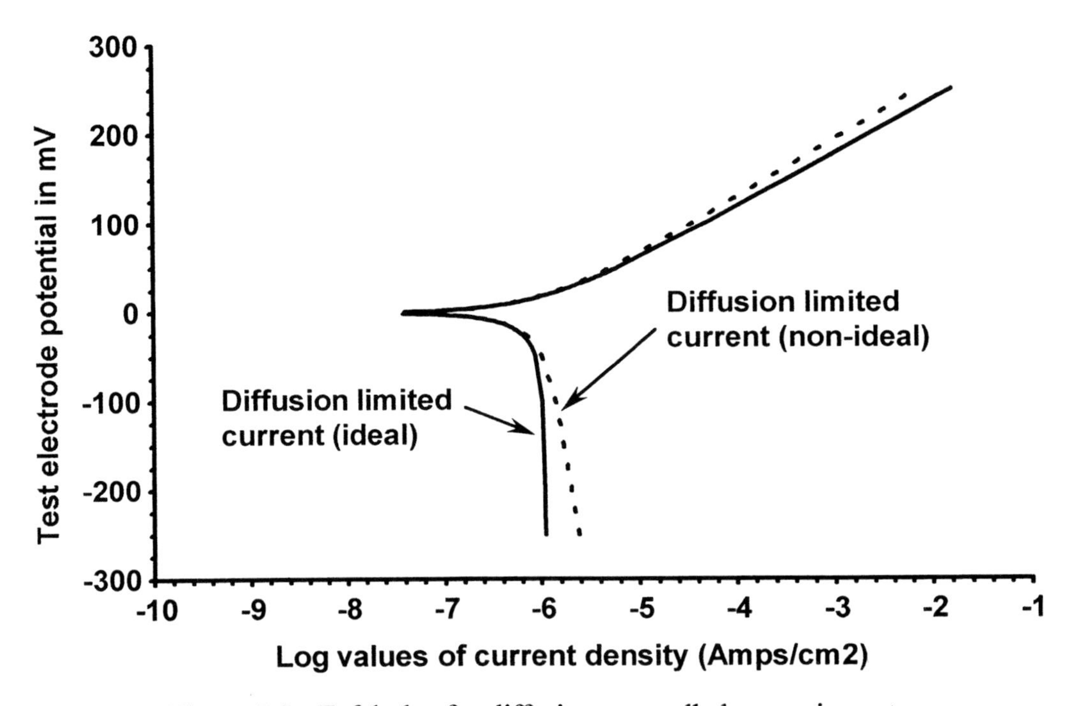

Figure 5.2. Tafel plot for diffusion controlled corrosion rates

Electrolyte stirring increases EAS transport to an electrode surface and increases the limiting current density magnitude as illustrated in Figure 5.3. Test electrodes in these examples were exposed to stagnant electrolyte (dashed plot) and a mildly-stirred electrolyte (solid plot). The diffusion limited current in Figure 5.3 is higher for the stirred solution. It appears in Figure 5.3 that stirring also changes OCP magnitude, but this is not the case because the approximately 75 mV difference between the two OCP values is within the 100 mV variation illustrated in Figure 2.7 (page 29).

It is important to verify that diffusion control is indeed present by comparing limited current magnitudes from different stirring rates, because other factors like a high scan rate can also produce a Tafel plot that looks like it is diffusion controlled. A Tafel plot is not diffusion controlled when increasing electrolyte stirring does not increase the cathodic curve current density.

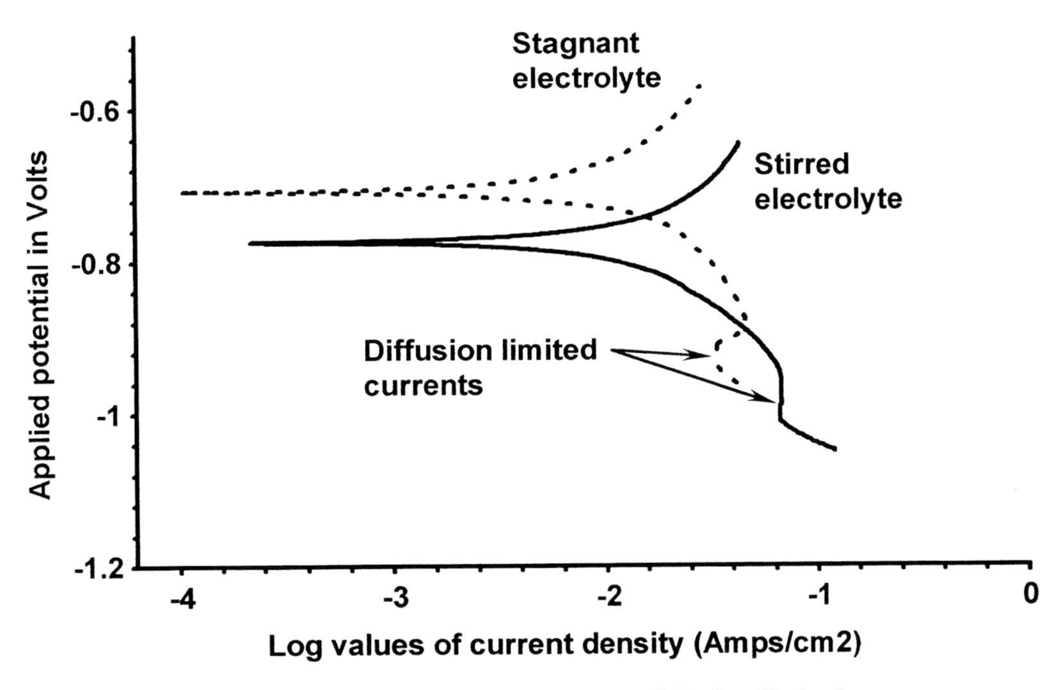

Figure 5.3. Electrolyte stirring increases diffusion limited current

5.5 Determining Tafel slopes

Tafel slopes have units of Volts per current-density-decade, where a decade is one order of magnitude current density, such as from 0.10 amp/cm^2 to 1.0 amp/cm^2. Figure 5.4 illustrates how to estimate the cathodic Tafel slope (β_c) for the Figure 5.1. Tafel plot.

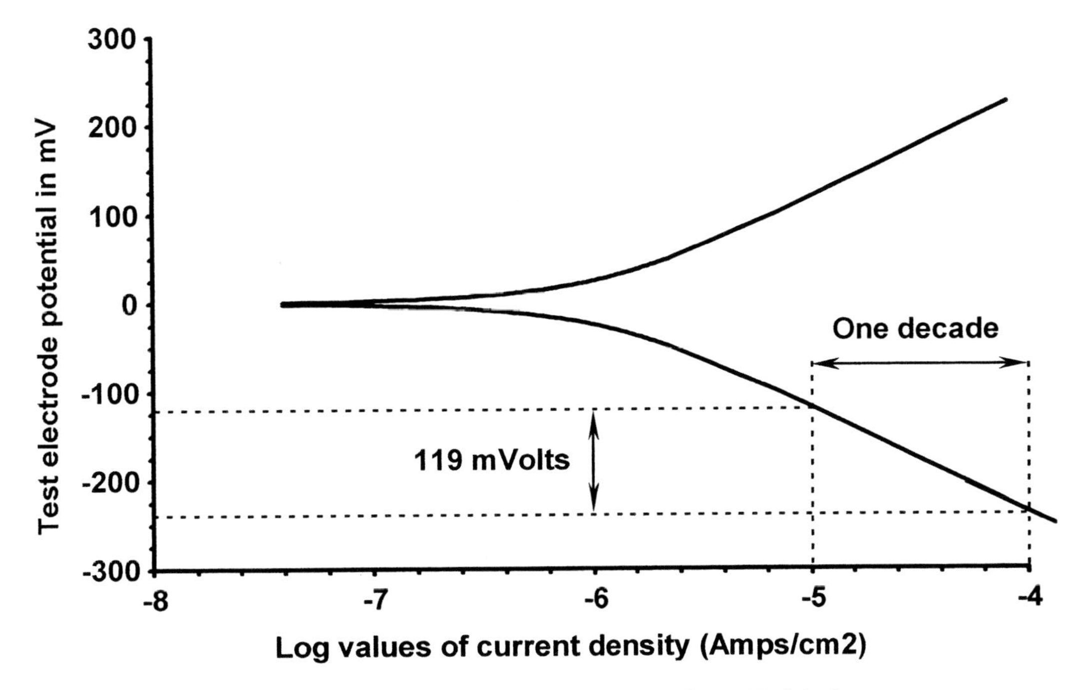

Figure 5.4. Determining Tafel slopes from Tafel plots

In this example the cathodic Tafel slope, β_c, is:

$$\beta_c = (0.119 \text{ Volts})/(1 \text{ decade})$$
$$= 0.119 \text{ Volts/decade}$$
$$= 119 \text{ mV/decade}$$

Tafel slopes can be used with linear polarization data when more accurate corrosion rates are desired.

5.6 Corrosion rates from Tafel plots

Corrosion current is read directly from a Tafel plot without the need for Tafel slope values or use of the Stern Geary equation. Figures 5.5 and 5.6 illustrate how corrosion rates are obtained from both activation and diffusion controlled Tafel plots, respectively.

Corrosion current for an activation controlled Tafel plot is the intersection of the anodic and cathodic linear extrapolations at OCP, as shown in Figure 5.5. The Figure 5.5 corrosion current is 1.02×10^{-7} Amps/cm^2, which can be converted to a corrosion rate with Equation 4.3 (pages 48 and 49).

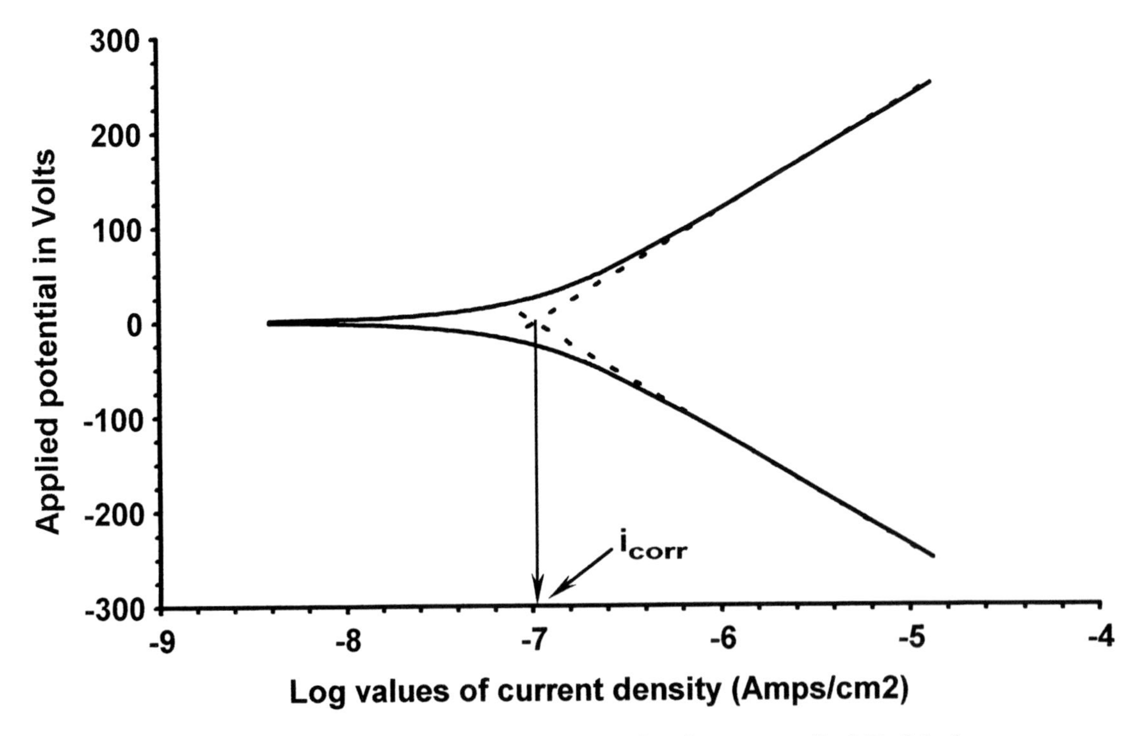

Figure 5.5. Corrosion current from activation controlled Tafel plots

Corrosion current is the diffusion limited current when cathodic current is constant with decreasing potentials as shown for the left plot in Figure 5.6 (ideal situation). The corrosion current in this example is $1x10^{-7}$ Amps/cm^2.

Corrosion current is **estimated** for a non-ideal diffusion controlled Tafel plot. The intersection of linear extrapolations for both branches at OCP is an **estimated** corrosion current, as shown for the right curve in Figure 5.6. Corrosion current for the right plot (non-ideal) is **approximately** $8.9x10^{-6}$ Amps/cm^2.

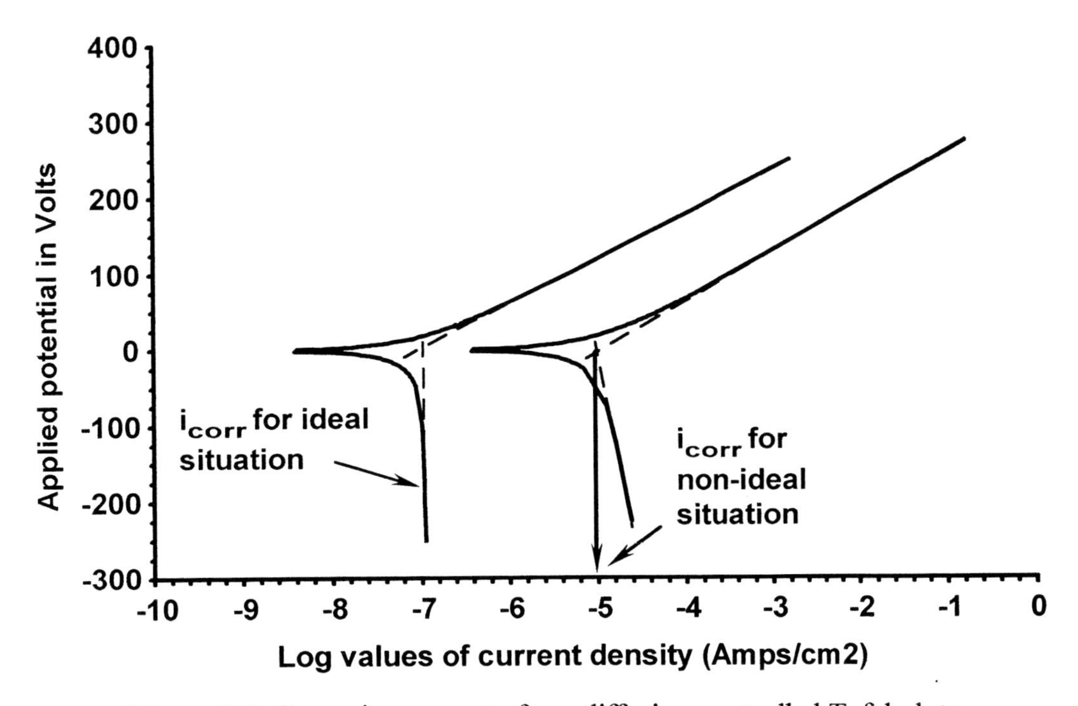

Figure 5.6. Corrosion currents from diffusion controlled Tafel plots

It is important to remember that diffusion limited current densities are a function of electrolyte stirring rate. Consequently, Tafel plot measurements should employ the hydrodynamic conditions (e.g., Reynolds number[4]) for the corresponding commercial system, if the measurement objective is to determine (or estimate) the corrosion rate of a commercial system, where electrolyte moves past metal structures or vessel walls. Unfortunately, transport equations are not well defined for electrolytes moving past vessel walls or complex shapes, making it difficult to reproduce commercial hydrodynamic conditions in Tafel plot corrosion measurements.

5.7 Tafel plot errors

Corrosion data may not correspond to actual corrosion rates when electrolyte solution resistance is high. For example, Mansfeld demonstrated that measured currents were significantly lower than actual currents when Tafel plot data were not corrected for uncompensated solution resistance.[2] The Butler Volmer equation will be used to quantitatively illustrate how uncompensated solution resistance can affect Tafel plot current magnitude and how uncompensated current compares with that for a corrected Tafel plot. Remember that the Butler Volmer equation is:

$$i = i_{corr}\left[\exp\text{-}(\alpha nF\eta/RT) - \exp(1-\alpha)nF\eta/RT)\right] \qquad [5.1]$$

We will use:

- $n = 2$
- $\alpha = 0.3$
- $i_{corr} = 1\text{x}10^{-6}$ Amps/cm^2
- IR-drop = 40 mV
- a correct polarization voltage, η, of 50 mV from OCP
- nF/RT = to 0.078 mV^{-1} (T = 25°C)

Anodic currents have negative values in this example because of physics sign conventions for electrical voltage.[6] The calculation for currents with and without correction for uncompensated solution resistance are:

A. Tafel plot is corrected for uncompensated solution resistance

$$\begin{aligned} i\,(\text{Amps/ cm}^2) &= 1\text{x}10^{-6}\left[\exp\text{-}[(0.3)(0.078)(50)] - \exp[(0.7)\,(0.078)(50)]\right] \\ &= 1\text{x}10^{-6}\left[\exp(-1.17) - \exp(2.73)\right] \\ &= 1\text{x}10^{-6}\left[0.3104 - 15.33\right] \\ &= -1.50\text{x}10^{-5}\ \text{Amps/cm}^2 \end{aligned}$$

B. Tafel plot is not corrected for uncompensated solution resistance
(remember that the IR-drop is 40 mV, so η = 50 - 40 = 10 mV because potential is not corrected for uncompensated solution resistance)

$$\begin{aligned} i\,(\text{Amps/ cm}^2) &= 1\text{x}10^{-6}\left[\exp\text{-}[(0.3)(0.078)(10)] - \exp[(0.7)\,(0.078)(10)]\right] \\ &= 1\text{x}10^{-6}\left[\exp(-.234) - \exp(.702)\right] \\ &= 1\text{x}10^{-6}\left[0.791 - 2.02\right] \\ &= -0.123\text{x}10^{-5}\ \text{Amps/cm}^2 \end{aligned}$$

The percentage difference between these two currents is:

$$\Delta \text{ (as percent)} = \left[(i_{corrected} - i_{uncorrected})/(i_{corrected})\right] \text{x}100\%$$
$$= \left[-1.50\text{x}10^{-5} - 0.123\text{x}10^{-5}\right]/\left[-1.50\text{x}10^{-5}\right]\text{x}100\%$$
$$= 91.8\%$$

It can be seen from this example that the 40 mV IR-drop causes an approximately 92% difference between uncorrected and actual currents. Tafel plot measurements should probably incorporate the current interrupt method when the IR-drop is expected to be greater than one millivolt. Electrochemical impedance spectroscopy (discussed in Chapters 7 and 8) can be used to measure uncompensated solution resistance and thus to estimate IR-drop magnitude prior to conducting Tafel plot measurements.

Asymmetrical electrode shapes (see pages 25 through 28) and measurements made prior to steady state (see pages 38 through 42) can also cause Tafel plot errors. Tafel plot scan rate can also cause errors, particularly when potentials change before electrical double layer chemistry has adjusted to each potential.

5.8 Limitations

A test electrode can be polarized only a limited number of times when making Tafel plot measurements, because some degree of electrode surface roughening occurs with each polarization. There are no hard-fast rules for determining when repeated polarizations have altered the test electrode surface to the point where a new test electrode should be used for further measurements. Figure 2.7 on page 29 demonstrates that OCP values can vary over a range of approximately 100 mV for electrodes that have not been polarized. It is recommended to replace a test electrode when the difference between its original and subsequent OCP values exceeds 100 mV, when the actual OCP variability for the electrolyte-metal system is unknown.

Tafel plots can not be used to determine whether or not a metal-electrolyte system is passive (no corrosion occurs), nor can Tafel plots be used to study or measure localized corrosion. Much larger anodic polarizations are required to obtain these types of information.

5.9 Summary

Tafel plots can be used to: a) estimate Tafel slopes, b) estimate corrosion rates and c) determine what type of chemical process controls corrosion rates.

A test electrode can be polarized only a limited number of times, because each Tafel plot polarization causes some degree of surface roughening. Test electrode OCP can be used to determine when to replace a test electrode. A 100 mV difference between original and subsequent test electrode OCP values can be used as a criterion for test electrode replacement, when actual OCP variability is unknown for a metal-electrolyte system.

Tafel plots are also subject to errors that arise from uncompensated solution resistance and asymmetrical test electrode shape.

The next chapter discusses larger voltage polarizations and how their use increases the amount of corrosion information that can be obtained from DC electrochemical corrosion measurement methods.

5.10 Reference

1) A. J. Bard and L. R. Faulkner, Electrochemical Methods: Fundamentals and Applications, John Wiley & Sons, NY (1980)
2) F. Mansfeld, Electrochemical Techniques, edited by R. Baboian, p. 69, National Association of Corrosion Engineers, Houston, TX (1986)
3) W. S. Tait, Corrosion, **35** (7), pp. 296-300 (1979)
4) W. L. McCabe and J. C. Smith, Unit Operations of Chemical Engineering, third edition, pp. 51-52, McGraw-Hill Book Company, New York (1976)
5) H. Kaesche, Metallic Corrosion, p. 92, National Association of Corrosion Engineers, Houston TX (1985)
6) F. W. Sears and M. W. Zemansky, University Physics, third edition, pp. 623-635, Addison-Wesley, Reading, MA (1964)

CHAPTER 6

Wide Potential Range Direct Current Polarizations

Objectives

After completing this Chapter, you will understand:

- that wider polarization potential ranges provide more information about corrosion
- how to generate potentiodynamic scanning (PDS) and cyclic polarization (CP) direct current polarization curves
- how to use PDS and CP curve features to characterize and predict long-term corrosion
- how potential scan rate affects PDS and CP curve features
- how to recognize positive and negative hysteresis in CP curves
- how pitting rates have been estimated from CP curves for certain situations
- the limitations of PDS and CP corrosion measurement methods

6.1 Introduction

The previous two chapters illustrated how the amount of information obtained from electrochemical corrosion measurements increased, as wider polarization potential ranges are used to generate the data. For example, linear polarization provides general corrosion rate information with a 20 to 40 mV potential range; Tafel plots provide corrosion rate and kinetic information with a 400 to 500 mV potential range. This chapter will complete discussion on direct current electrochemical measurement methods with discussion of Potentiodynamic scanning (PDS) and Cyclic Polarization (CP) curves. These types of curves are generated with approximately 1250 mV to 2250 mV potential ranges, respectively, and provide additional information about corrosion kinetics and localized corrosion.

6.2 More Corrosion Terms

New corrosion terms are defined on the next page and throughout this chapter. Many of the terms presented in the first five chapters will also be used, so You may want to review corrosion terminology defined in previous chapters. Chapter 1 definitions are on pages 2 and 3; Chapter 2 on pages 18 and 19; Chapter 3 on pages 37 and 38; Chapter 4 on page 44; and Chapter 5 on pages 53 and 54.

Active corrosion behavior:

Active corrosion behavior is typically observed when a metal produces visible quantities of corrosion after brief exposure to an electrolyte (e.g., twenty four hours exposure). Visible corrosion is typically a porous hydroxide layer that adheres loosely to the metal and does not provide very good corrosion protection. General and pitting corrosion often occur together when a metal exhibits active corrosion behavior.

Passive corrosion behavior:

Passive corrosion behavior occurs when a thin protective (passive) film forms on a metal surface. Corrosion either does not occur, or the corrosion rate is so low that it does not significantly reduce service lifetime, when a passive film forms. The exact structure of a passive film is unknown, but it is hypothesized that passive films are 20 to 100 Angstrom thick metal oxides (1 Angstrom is 10^{-8} cm).

6.3 Potentiodynamic scanning (PDS) and cyclic polarization (CP) curve structures

Potentiodynamic scanning (PDS) and cyclic polarization (CP) curves have a cathodic branch that is similar to that for a Tafel Plot. PDS and CP curves also have anodic branches, but they extend over a wider potential range and are often much more complex than Tafel plot anodic branches. This chapter focuses on the additional information obtained when anodic branch polarization is extended 700 to 800 mV beyond the range used for Tafel plots. Descriptions of PDS and CP curves will be separated in this section, but other sections will combine discussions on these two curve types.

A. Potentiodynamic scanning curves

Figure 6.1 contains an example PDS curve. Comparison of the Figure 6.1 PDS curve with the Figure 5.1 Tafel plot (page 55) shows that several additional quantities appear in the PDS anodic branch:[1]

1. the primary passivation potential, E_{pp}, is the potential after which current either decreases, or becomes essentially constant over a finite potential range
2. the breakdown potential, E_b, is the potential where current increases with increasing potential
3. the passive region is the portion of the curve between E_{pp} and E_b

4. the portion of the PDS curve where potentials are less (more negative) than E_{pp} is referred to as the active region of the curve
5. the portion of the curve where potentials are greater (more positive) than E_b is referred to as the transpassive region of the curve

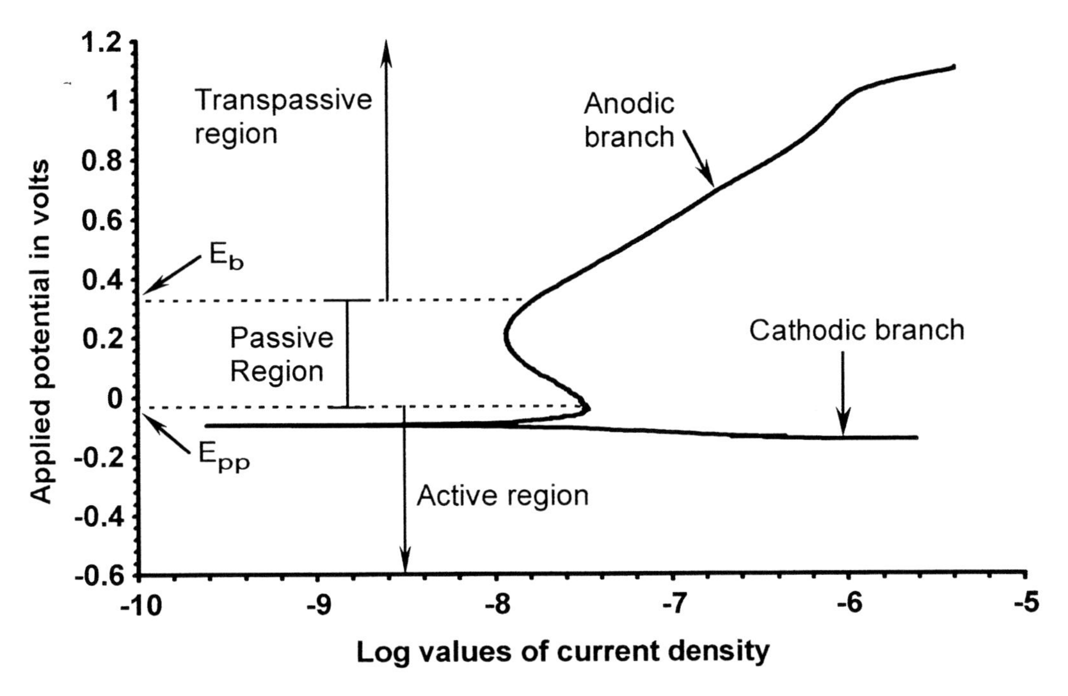

Figure 6.1. PDS curve for passive corrosion behavior

Quantities like E_{pp}, E_b, and passive region width can be used to characterize corrosion behavior, and evaluate how effectively a passive film protects a metal from corrosion. General corrosion, and sometimes pitting, occurs in the active region; little or no corrosion occurs in the passive region; and pitting corrosion can occur in the transpassive region.[2]

B. Cyclic polarization curves

Cyclic polarization curves can be considered as extensions of PDS curves. Test electrode potential is increased in the anodic direction (for the anodic branch), until test electrode polarization either reaches approximately +1000 mV from OCP, or current density reaches a given magnitude, then potential is decreased toward open circuit potential (OCP). Figures 6.2 and 6.3 contain examples of CP curves. The dashed arrows next to the forward and reverse anodic branches indicate potential scan directions.

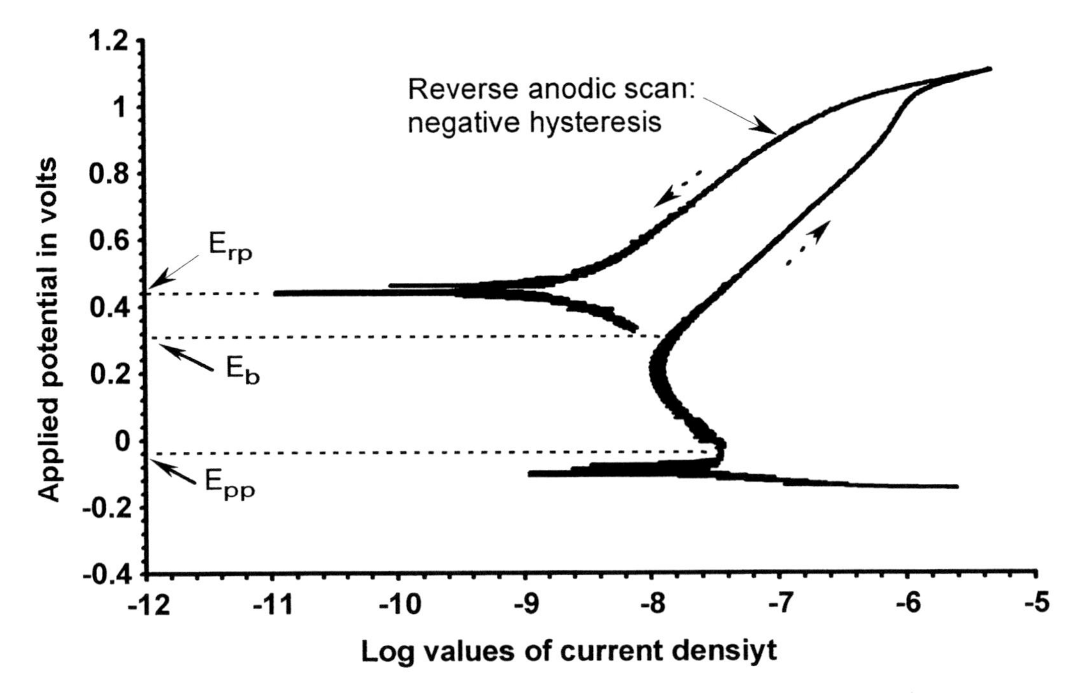

Figure 6.2. CP curve with negative hysteresis

Three new quantities, in addition to those identified for PDS curves in Figure 6.1, can be observed in a CP curve:

a) negative hysteresis (Figure 6.2), or
b) positive hysteresis (Figure 6.3)
c) repassivation potential, E_{rp}

Negative hysteresis (Figure 6.2) occurs when reverse scan current density is less than that for the forward scan, and positive hysteresis (Figure 6.3) occurs when reverse scan current density is greater than that for the forward scan.

A passive film is damaged when potential is raised into the transpassive region of a PDS or CP curve, and pits can initiate when film damage is at discrete (localized) locations on the metal surface.[3,4] It is generally believed that pits will continue to grow when OCP is greater than E_{rp}, and pits will not grow when OCP is less than E_{rp}.[5,6]

CP curve hysteresis can provide information on pitting corrosion rates and how readily a passive film repairs itself. Positive hysteresis occurs when passive film damage is not repaired and/or pits initiate; negative hysteresis occurs when a damaged passive film repairs itself and pits do not initiate.[7,8] Section 6.5 contains a discussion on a case history where CP curve hysteresis was also used to estimate pitting corrosion rates for non-crevice areas. Other case histories will be also be discussed in section 6.5 to illustrate how PDS and CP methods can be applied to industrial corrosion problems.

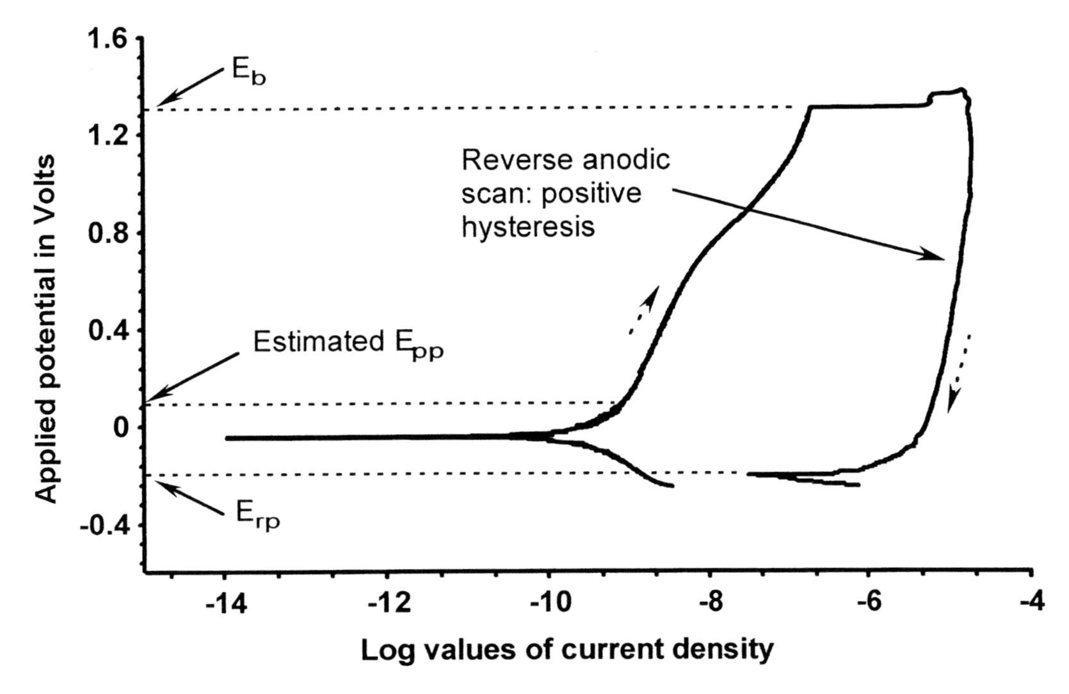

Figure 6.3. CP curve with positive hysteresis

6.4 Generating PDS and CP curves

One or two test electrodes can be used to generate PDS and CP curves.

Single test electrode method

Anodic and cathodic branches are generated from the same test electrode. Polarization is begun for both PDS and CP curves at approximately -200 to -250 mV from OCP and increased until the potential is approximately +1000 mV from OCP. Anodic polarization for CP curves is reversed at either approximately +1000 mV from OCP, or at a specified current density and the scan proceeds back to OCP.

Two test electrode method

Separate test electrodes are used in the same electrolyte to generate anodic and cathodic branches. The cathodic branch is generated by polarizing one of the test electrodes from OCP to approximately -200 to -250 mV from OCP, and the anodic branch is generated by polarizing the other test electrode from OCP to approximately +1000 mV from OCP for PDS curves. Anodic polarization for CP curves is reversed at either approximately +1000 mV from OCP, or at a specified current density, and the scan proceeds back to OCP.

The original OCP (OCP measured prior to polarization) is the starting (or reference) point for PDS and CP measurements, as it is with the other DC methods discussed in Chapters 4 and 5. Applied potential is typically increased (or decreased) in 1 to 2 mV steps.

The two electrode method can produce PDS and CP curves with a gap between the anodic and cathodic branches, particularly when OCP values vary 100 mV or more as shown in Figure 2.7, on page 29 of Chapter 2.

The single electrode method can produce PDS and CP curves in which the original OCP is not located at the inflection point between the anodic and cathodic branches (potential where current is zero), but is located in the passive region of the curve as shown by the solid curve in Figure 6.4. This situation is typically observed when the single electrode method is used for metals that spontaneously passivate, or form barrier oxide films on their surface.

A passive film or barrier oxide is part of a test electrode electrical double layer. Remember that changing double layer chemistry changes test electrode potential (e.g., Equation 1.8 on page 10). Beginning PDS or CP polarizations at -200 to -250 mV from OCP causes some reduction of metal ions in a barrier or passive oxide film,[9] so it is reasonable to conclude that metal ion reduction changes test electrode potential. A passive film or barrier oxide begins returning to its natural chemistry and crystalline structure as polarization approaches the potential where current is zero (inflection point in Figure 6.4). This inflection occurs because potentials around this point are no longer at a magnitude that causes ion reduction, metal ions begin re-oxidizing, and the film or oxide begins to re-crystallize. However, film chemistry and crystal structure do not completely return to their original conditions at the inflection point, and the curve is shifted toward negative potentials as illustrated in Figure 6.4, resulting in OCP being located in the passive region. The steep cathodic branch slope (for the solid curve in Figure 6.4) indicates that barrier oxide or passive film chemistry changes occur rapidly during initial stages of polarization. Oxide or film chemistry changes are less rapid as potentials approach E_{pp}; the majority of film and oxide changes that occur after E_{pp} (during polarization in the passive region) are due to increasing film or oxide thickness and re-crystallization.[10]

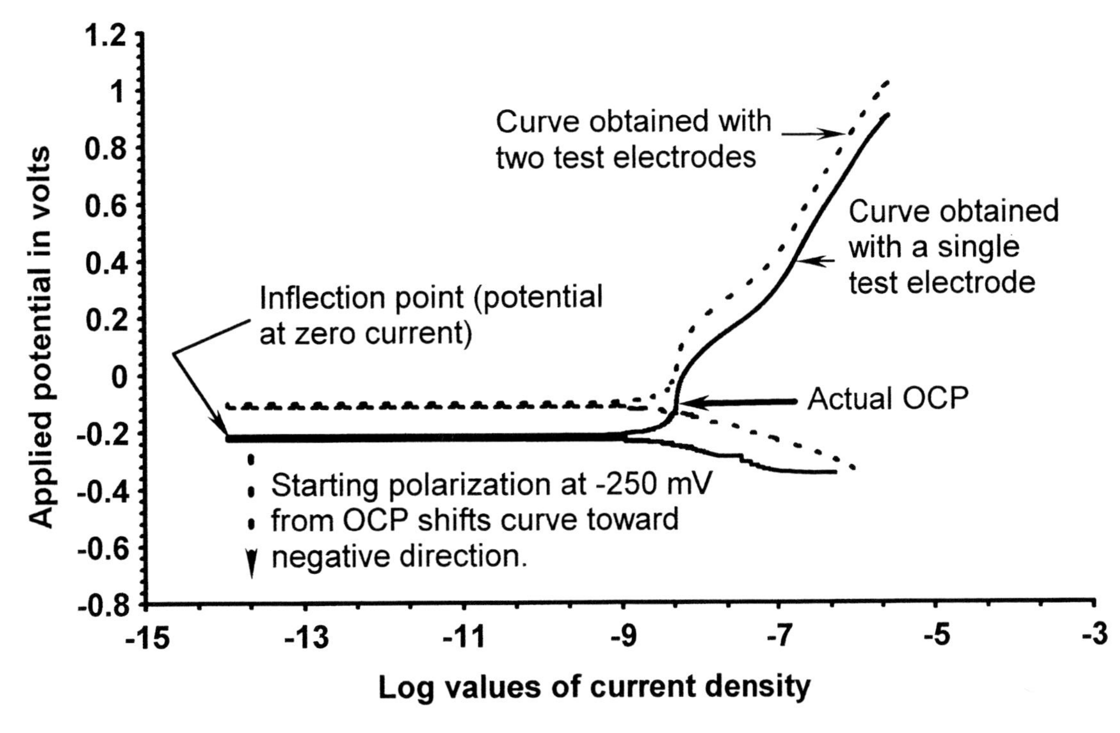

Figure 6.4. Situation where OCP is not equal to potential at PDS curve inflection

6.5 How scan rate affects PDS and CP curves

E_b, E_{pp} and E_{rp} are not intrinsic properties of a metal-electrolyte system because their magnitudes are dependent upon scan rate.[11,12,13,14] E_{rp} magnitudes also have been shown to be determined by extent of pit growth.[15,16,17] Consequently, passive region width (E_b- E_{pp}) is not an intrinsic property. However, these quantities can be used to qualitatively: a) characterize passive behavior stability, b) evaluate affects of various inhibitors on metal corrosion in a given electrolyte, and c) evaluate different metals for a given electrolyte.

Not all PDS and CP curves have the shape shown in Figures 6.1 and 6.2. Some curves have less distinct E_b and E_{pp} potentials (like Figures 6.3 and 6.4); some have passive regions where current increases slightly as potential increases (like Figure 6.3); some have OCP values that coincide with E_b; some have OCP values in the active region; and some have hysteresis that appears to be both positive and negative.

6.6 Case histories: use of PDS and CP curves for industrial corrosion problems

This section discusses four industrial case histories where PDS or CP curves were used to evaluate long-term corrosion behavior. These cases are included to illustrate the usefulness of direct current polarization corrosion measurement in industrial situations, but do not represent all possible types of PDS and CP curves that You may encounter. Interpretations in this section may not apply to industrial corrosion situations that are significantly different from conditions described with each case. All curves in this section were generated using the single electrode method (same test electrode used to generate both curve branches).

Case History number 1:

Long-term passive corrosion behavior where the CP curve has distinct E_b and E_{pp} potentials, and OCP is in the passive region

Figure 6.5 contains a PDS curve for steel metal containers exposed to an inhibited, pH 8, aqueous electrolyte. Notice that

a) there are distinct E_b and E_{pp} potentials and a passive region

b) OCP is located in the passive region

c) the reverse scan has negative hysteresis.

This curve is interpreted as indicating passive steel corrosion behavior that will provide at least two year container service lifetime. Pitting corrosion is not expected to occur because hysteresis is negative.

Actual corrosion corroborated this interpretation because no pitting and only slight steel crevice corrosion were observed after two years exposure in this case.[18] This type of curve and interpretation has also been successfully applied to stainless steel processing vessels, as well as brass and aluminum exposed to a variety of other electrolytes.

It should be mentioned that OCP for a metal is an average value for the entire exposed metal surface, and there can be local areas on the metal surface, as well as in crevices that have potentials different from OCP.[19,20,21] Consequently, there can be isolated areas on the surface or in crevices that have potentials in the active or transpassive regions, particularly when the OCP is within 100 millivolts of E_b or E_{pp}, or the passive region width is narrow (e.g., only 100 to 200 mV wide). Areas that have potentials in: a) the active region would be expected to have general corrosion, and possibly pitting corrosion, and b) the transpassive region would be expected to have pitting corrosion. Occurrence of general crevice corrosion without pitting for this case history was probably due to crevice OCP being located in the PDS curve active region.

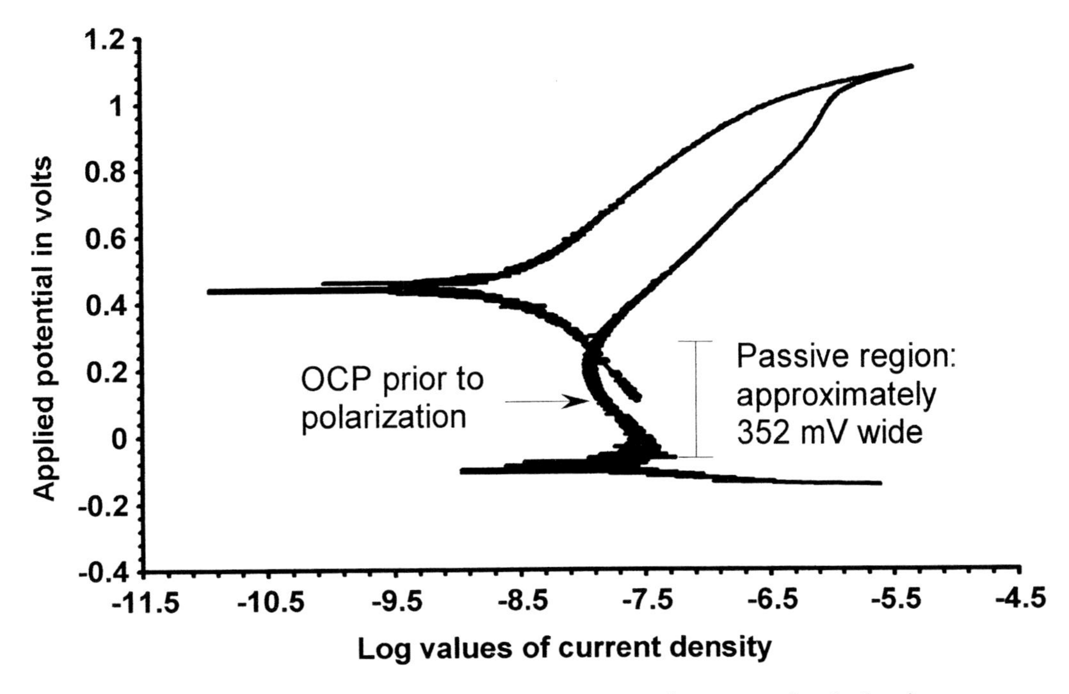

Figure 6.5. PDS curve for long-term passive corrosion behavior

There are occasions when E_{pp} does not have a distinct maximum current like those in Figures 6.1 and 6.5, but the metal spontaneously passivates.

<u>Case History number 2:</u>

Long-term passive corrosion behavior where E_b and E_{pp} potentials can only be estimated

Figure 6.6, on the next page, was obtained from type 1018 mild steel in the water phase from a broken water-in-oil emulsion. The curve does not have a distinct E_{pp} potential, but there appears to be an E_b potential and OCP is located in an area where current does not increase as potential increases. Hysteresis is also negative for this curve.

This type of curve is referred to as a pseudo-passive curve because:

- there is no distinct E_{pp} potential
- there appears to be a transpassive region and an E_b potential
- OCP is in an area where current does not increase with potential

Interpretation for Figure 6.6 is the same as for a passive curve like that in the previous case.

No steel general or pitting corrosion was observed after three years exposure in this case, corroborating the interpretation for Figure 6.6.

One method for qualitatively estimating E_b and E_{pp} potentials is to extrapolate the linear portions of the active, passive and transpassive regions as shown by the dashed lines in Figure 6.6. The difference between E_b and E_{pp} is the estimated passive region width. E_b is approximately 135 mV in Figure 6.8, E_{pp} is approximately -33 mV, and the passive region width is approximately 168 mV.

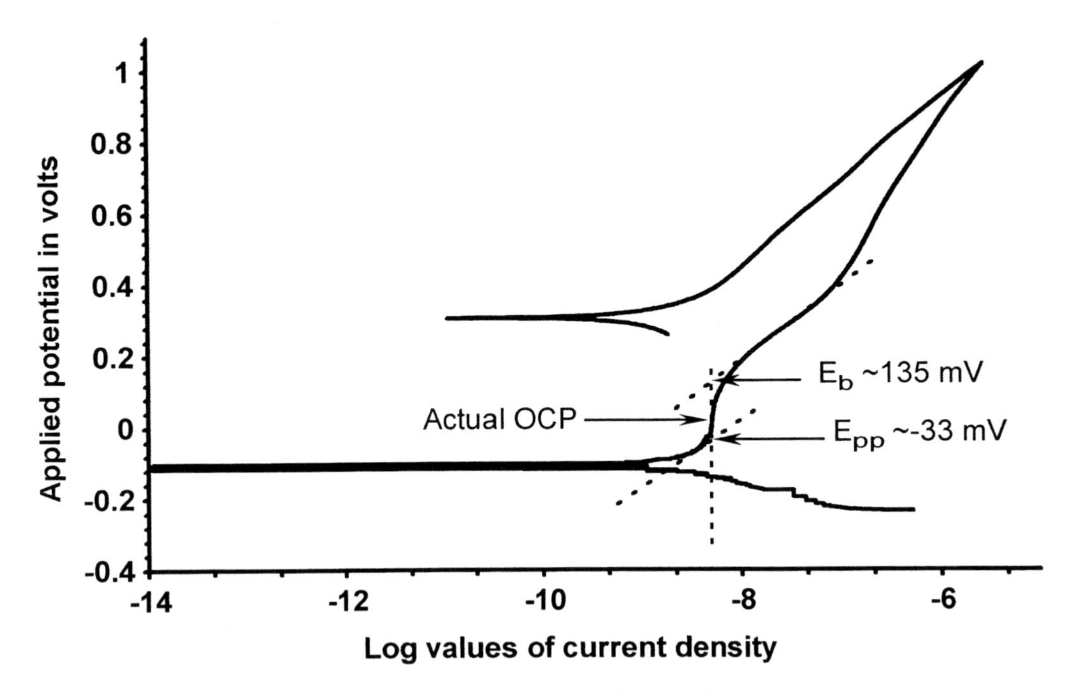

Figure 6.6. Example of a pseudo-passive CP curve

It should be remembered that E_b, E_{pp} and passive region width values for a pseudo-passive curve are only estimates. It is often difficult to precisely choose where to begin extrapolations from the active, passive and transpassive regions.

There are also cases where the current may increase slightly in a pseudo-passive region such as occurs in Figure 6.3 (page 67). Passive region slope is not an intrinsic property of a metal-electrolyte system because scan rate can affect slope magnitude and even obliterate a passive region when an excessively fast scan rate is used.[22] Consequently, it is possible that a pseudo-passive CP or PDS curve is actually a true passive curve generated with an excessively fast potential scan rate. However, pseudo-passive CP curves can be used to: a) determine if a metal passivates or forms a barrier oxide, b) estimate passive region width, and c) qualitatively compare corrosion behavior for different metals and electrolytes.

In summary, a pseudo-passive PDS or CP curve has a transpassive and passive region with OCP in the pseudo-passive region, but may not have distinct E_b or E_{pp} potentials. Current may increase slightly in the pseudo-passive region, but slope magnitude is not an intrinsic property of a metal-electrolyte system. The passive region for pseudo-passive curves (like passive curves) may not be wide enough to prevent corrosion from occurring at local areas or in crevices. Also, estimation of E_b and E_{pp} pseudo-passive curve values can be difficult and thus somewhat subjective.

Case History number 3:

Active corrosion with high general and pitting corrosion rates

Figure 6.7 contains an example of an active PDS curve for type 1018 mild steel in oxygen-saturated water. [7] No E_{pp} or E_b potentials appear to be present and current continues increasing with potential throughout the entire forward anodic potential scan.

This curve is interpreted as indicating that OCP, E_{pp} and E_b values are equal, or separated by less than 100 millivolts, thus general and pitting corrosion are expected to occur.

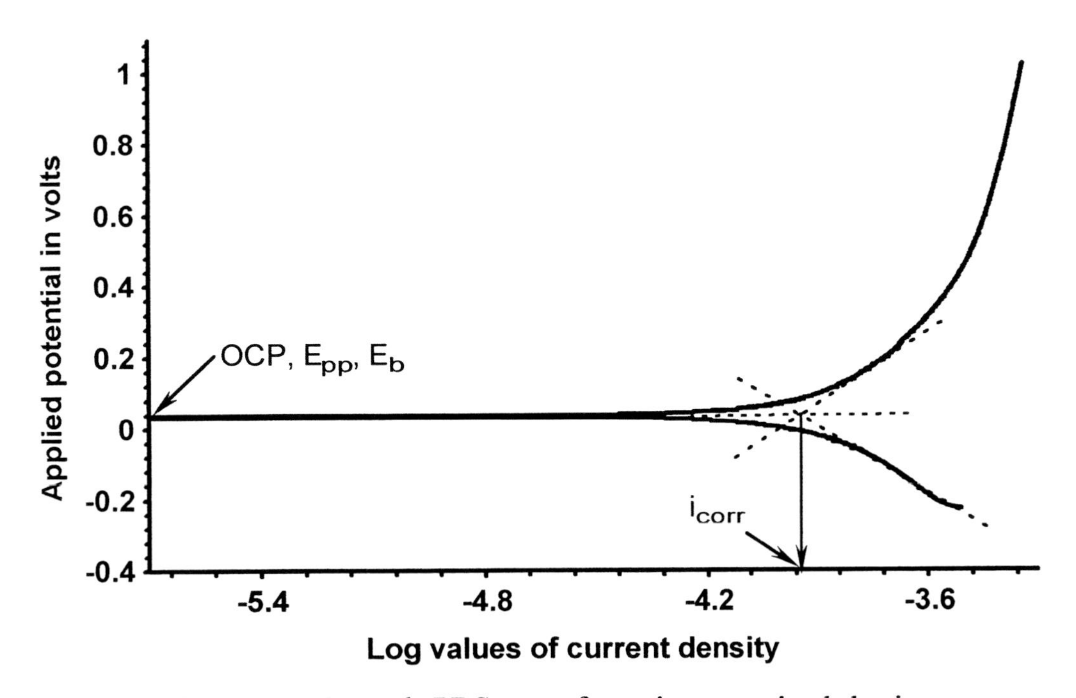

Figure 6.7. Example PDS curve for active corrosion behavior

Mild steel had approximately 50 mpy general and 500 mpy (1/2 inch per year) pitting corrosion rates, corroborating curve interpretation in this case.[7]

General corrosion rates can be estimated from active CP and PDS curves by extrapolating the linear portions of the anodic and cathodic branches to OCP (as shown by dashed lines in Figure 6.7), similar to estimating i_{corr} from Tafel plots (see pages 58 and 59). However, Tafel plots give much more accurate estimations for i_{corr}, because it is often difficult to determine where is the linear portion of PDS and CP anodic and cathodic branches. Several potentiostat manufacturers provide software that estimates corrosion rates from PDS curves.

In summary, PDS and CP curves are considered to be active when current continues increasing (with increasing potential) over the entire anodic branch forward scan. General corrosion and pitting may both occur, particularly when E_b coincides with the natural OCP.

Case History number 4:

Estimating pitting rates from reverse anodic scans

CP curves exhibit positive hysteresis when applied potential is extended into the transpassive region and the passive film is unable to repair breakdown at localized areas.[23] Negative hysteresis occurs when a passive film is able to repair itself as potential is decreased. Consequently, Hysteresis can be used to qualitatively evaluate pitting tendency of a metal in a given electrolyte.

The reverse CP anodic scan has been used to estimate pitting current density, i_{pit}, for admiralty brass and type 1018 mild steel in oxygen-saturated cooling water.[7] Figure 6.8 demonstrates how pitting currents can be estimated by extrapolating the reverse anodic scan to an intersection with the corresponding CP curve cathodic branch. This curve was obtained for stainless steel exposed to an inhibited, low chloride, pH 4 aqueous solution.

The current density at the intersection in Figure 6.8 is converted to a pitting rate using equation 4.3 (page 48), multiplied by a proportionality constant. For example, in Figure 6.8 the intersection of the reverse anodic scan and cathodic branch is approximately 1.5×10^{-9} Amps/cm^2. If we use a proportionality constant of 20 (see reference number 7), then the estimated pitting rate would be:

$$MPY_{pit} \cong 20(1.5 \times 10^{-9})(1.2866 \times 10^{5})(1/7.86)(27.56)$$

$\cong$ 0.014 mils per year

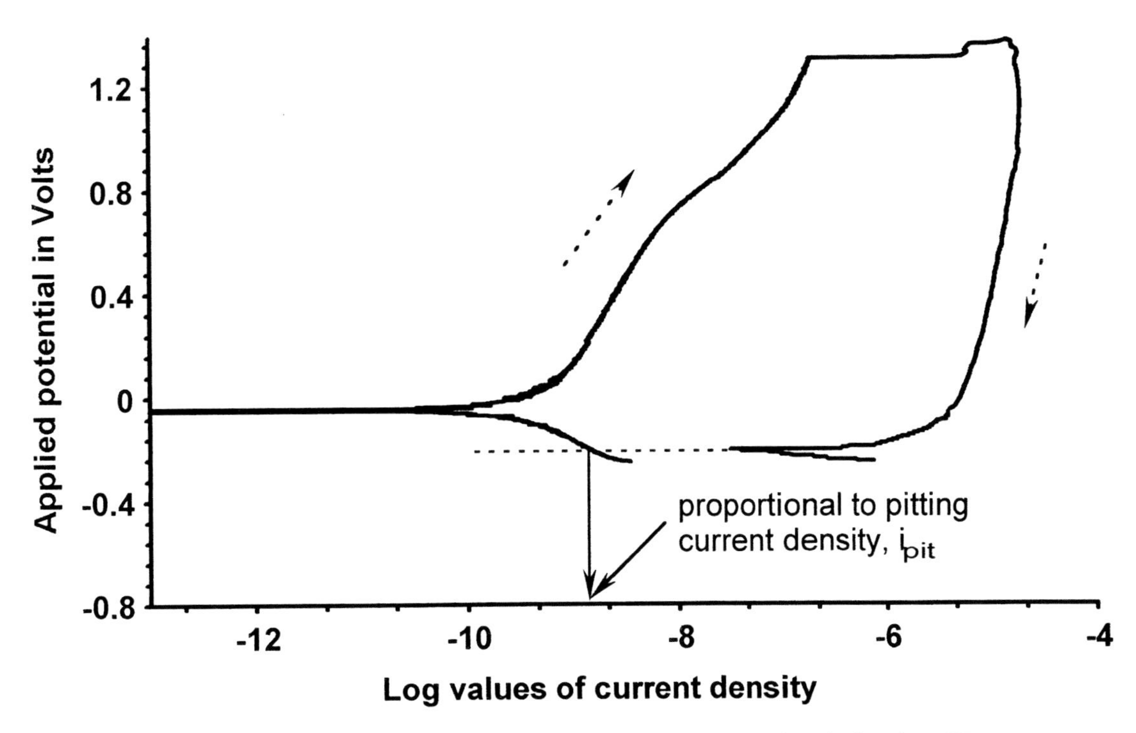

Figure 6.8. Estimating pitting rates from passive corrosion behavior CP curves

This pitting rate is a very low, leading to the conclusion that pitting will not be a serious problem, particularly if the vessel walls are thick (typical vessel wall thickness in this case was approximately 250 mils). In addition, the passive region width is approximately 1100 mV, which suggests that pitting is unlikely, even for areas whose potentials are not equal to OCP. Consequently, the interpretation for Figure 6.8 is that pitting of the vessel is unlikely, and pits will grow at a negligible rate in the unlikely event pitting does occur.

Actual experience qualitatively corroborated this interpretation because no pitting was observed in the vessel after four years of operation. The corroboration is considered to be qualitative in this case because the actual pitting rate for the vessel was never measured.

The pitting proportionality constant is only known for mild steel and admiralty brass in oxygen-saturated water. However, pitting current estimation can be used on other systems to qualitatively evaluate pitting resistance of different metals in the same electrolyte, or evaluate how different electrolytes affect pitting of a given metal, as long as you remember that this current is not the actual pitting current. Reference number 7 contains other examples on how pitting current is estimated from different types of CP curves from mild steel and admiralty brass in oxygen-saturated water.

It should be emphasized that this type of pitting rate estimation may not apply to all situations and is only for non-crevice areas. Crevice geometry plays a large role in determining when crevice corrosion occurs; thus it is difficult to measure or predict crevice corrosion with a non-crevice type electrode.[24]

6.7 Limitations of PDS and CP curves

CP and PDS anodic polarizations can cause high currents and pitting corrosion on test electrodes, which can alter electrode surface. Consequently, CP and PDS curves are typically generated only once on a given test electrode, particularly when a) positive hysteresis is observed in a CP curve, b) anodic currents are high (e.g., on the order of microAmps), or c) PDS anodic polarization extends into the transpassive region.

6.8 Summary

This chapter discussed CP and PDS polarization corrosion measurements and how to interpret several types of CP and PDS curves. Interpretation of all PDS and CP curve types was not attempted.

Corrosion behavior for a given curve is determined from: a) curve shape, b) OCP location in the curve, c) the presence or absence of a distinct E_{pp}, E_b and E_{rp} potentials and d) presence or absence of a passive region. Passive region width and slope, E_{pp}, E_b and E_{rp} potentials are not intrinsic properties of a metal-electrolyte system because their magnitudes can be influenced by scan rate and/or the extent of pitting corrosion on a metal surface.

6.9 References

1) A. J. Sedricks, Corrosion of Stainless Steels, pp. 54 - 57, J. Wiley & Sons, NY (1979)

2) G. R. Wallwork and B. Harris, Localized Corrosion, pp. 292 - 304, NACE, Houston, TX (1981)

3) Z. Szklarska-Smialowska, Pitting Corrosion of Metals, p. 40, NACE, Houston TX (1986)

4) Z. Szklarska-Smialowska, Op. Cit., p.4

5) Z. Szklarska-Smialowska, Op. Cit., pp. 39, 44-45

6) D. C. Silverman, Corrosion, **48** (9), p. 735 (1992)

7) W. S. Tait, Corrosion, **34** (6), pp. 214 - 217 (1978)

8) W. S. Tait, Corrosion, **35** (7), pp. 296 - 300 (1979)

9) Z. Szklarska-Smialowska, Op. Cit., p.12

10) J. P. Hoare, Passivity of Metals, p. 505, R. P. Frankenthal and J. Kruger, editors, Electrochemical Society monograph series, Princeton, NJ (1978)

11) Z. Szklarska-Smialowska, Op. Cit., p. 54

12) T. Yoshii and Y. Hisamatsu, J. Japan Inst. Metals, **36**, p. 750 (1992)

13) H. P. Leckie and H. H. Uhlig, J. Electrochem. Soc., **113**, p. 1261 (1966)

14) B. E. Wilde, Localized Corrosion, p. 346, NACE, Houston, TX (1981)

15) T. Suzuki and Y. Kitamura, Corrosion, **28** (1), p. 1 (1972)

16) G. Herbsleb and W. Schwenk, Corrosion Science., **13**, p. 739 (1973)

17) B. E. Wilde and E. Williams, Electrochem. Acta., **16**, p. 1971 (1971)

18) W. S. Tait and J. A. Maier, Corrosion, **42** (10), pp. 622 - 628 (1986)

19) D. Remppel and H. E. Exner, J. Electrochem. Soc., **138** (2), pp. 379 - 385 (1991)

20) J. Markworth, Corrosion, **47** (3), pp. 200 - 201 (1991)

21) B. G. Ateya and H. W. Pickering, J. Electrochem. Soc., **122** (8), pp. 1018 - 1026 (1975)

22) N. D.Greene and R. B. Leonard, Electrochemica Acta, **9**, p. 45-54 (1964)

23) R. L. Martin, Application of Electrochemical Polarization to Corrosion Problems, Petrolite Corp., St. Louis, MO (1977)

24) H. W. Pickering, 12th International Corrosion Congress, Houston TX (September 1993)

CHAPTER 7

Electrochemical Impedance Spectroscopy Fundamentals

Objectives:

After completing this Chapter, you will:

- understand how an electrified interface responds to an alternating current (AC) polarization
- understand how different AC polarizing voltage frequencies proceed through an electrified interface
- understand what chemical and physical processes produce different time constants
- be familiar with the three most common ways to graph EIS data
- understand the relationship between these three types of graphs

7.1 Introduction

Electrochemical impedance spectroscopy (EIS) uses a range of low magnitude polarizing voltages, much like linear polarization (Chapter 4). However, EIS voltages cycle from peak anodic to cathodic magnitudes (and vice versa) using a spectrum of alternating current (AC) voltage frequencies, instead of a range of single magnitude and polarity direct current (DC) voltages. Resistance and capacitance values are obtained from each frequency, and these quantities can provide information on corrosion behavior and rates, diffusion, and coating properties.

EIS measurements have been used on underground pipelines,[1] steel reinforcing bars in concrete,[2] uncoated metals exposed to chemical processing streams,[3] painted (coated) metals,[4] and internally coated metal containers.[5]

This chapter discusses principles underlying EIS. The next chapter discusses how to interpret and use data from EIS corrosion measurements. You may want to review pages 19 through 21 of Chapter 1, because the principles discussed in those pages also apply to EIS measurements.

7.2 Terminology related to EIS measurements

New terms are defined in this section and throughout the chapter. Terms introduced in previous chapters will also be used, so You may want to review corrosion terminology definitions in previous chapters. Chapter 1 definitions are on pages 2 and 3; Chapter 2 on pages 18 and 19; Chapter 3 on pages 37 and 38; Chapter 4 on page 44; Chapter 5 on pages 53 and 54; and Chapter 6 on page 64.

Capacitive reactance

Capacitive reactance is the ability of an electrified interface (e.g., an electrical double layer) to charge (and discharge) like an electrical capacitor in response to an external applied voltage.

Complex Impedance (commonly referred to as Impedance)

Impedance is the AC analogue of DC resistance. Impedance magnitude is equal to AC voltage magnitude divided by the corresponding AC current magnitude.

Diffusion

Diffusion is ion and molecule movement caused by concentration differences between different locations in an electrolyte. Diffusion direction is from higher to lower concentrations.

EIS Spectrum

An EIS spectrum is a series of AC voltage frequencies applied to a test electrode. Typical EIS spectra for corrosion measurements include frequencies from 100 kilohertz (100,000 hertz) to several millihertz (1 millihertz = 0.001 hertz).

Polarity

Polarity is the direction of a test electrode polarization produced by an external voltage. Test electrode polarity can be either anodic, or cathodic.

Time constant

A circuit time constant (in seconds) is equal to capacitance (in Farads) multiplied by the corresponding parallel resistance (in ohms).

7.3 AC and DC polarizations have different properties

A given DC voltage has constant magnitude, and its polarity (direction) is either anodic, or cathodic. An AC voltage cycles from peak anodic to peak cathodic amplitudes, as shown in Figure 7.1, and thus has variable magnitude with both anodic and cathodic polarity in each polarization cycle.

The AC voltage amplitude in Figure 7.1 is 5 mV (± 5 mV from OCP) and a complete cycle includes polarizing to a 5 mV peak anodic potential, discharging (to OCP = 0), polarizing to a -5 mV peak cathodic potential, discharging, and again polarizing to a 5 mV peak anodic potential. Two complete cycles are illustrated in Figure 7.1, thus the voltage frequency is 2 cycles/40 seconds, or 0.05 hertz (0.05 cycles per second).

EIS polarizing voltage amplitudes typically range from 5 to 20 millivolts and are usually centered around OCP as illustrated in Figure 7.1 (OCP = zero). Voltage frequencies used for EIS measurements can range from 100 kilohertz to several millihertz.

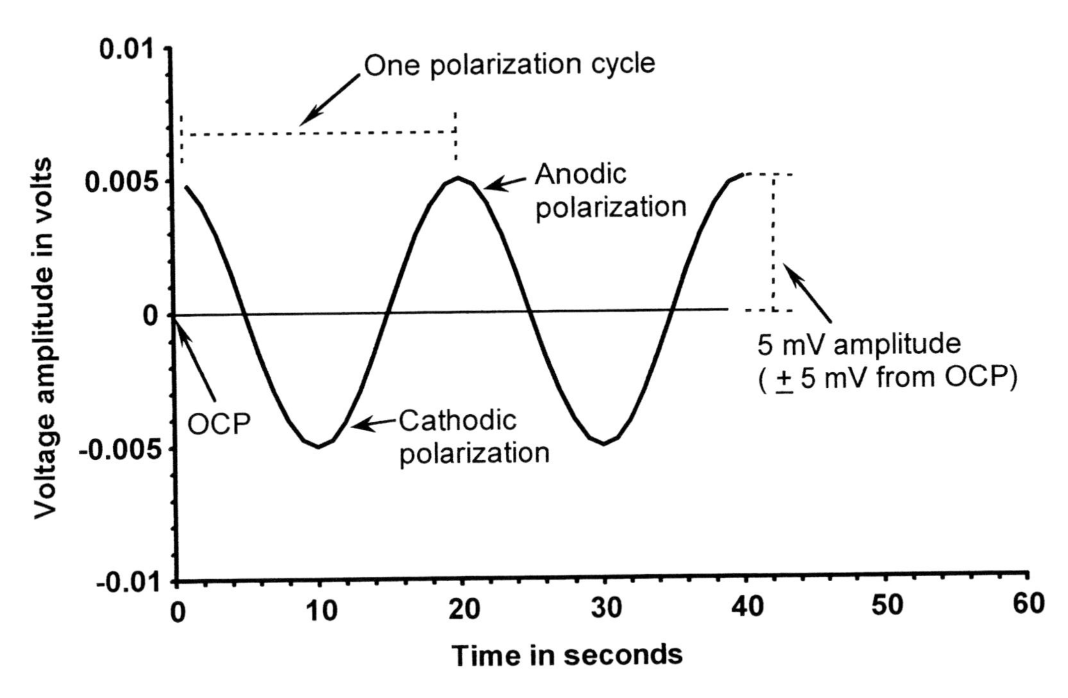

Figure 7.1. Cyclic nature of AC voltage

Degrees are equivalent to time in Figure 7.1, because degrees are equal to time multiplied by 360 and voltage frequency.

7.4 AC polarization phase behavior

An EDL can have electrical properties similar to those for a simple electrical circuit composed of resistors and a capacitor such as that in Figure 7.2 (see also pages 7 through 9 and 22 through 24).

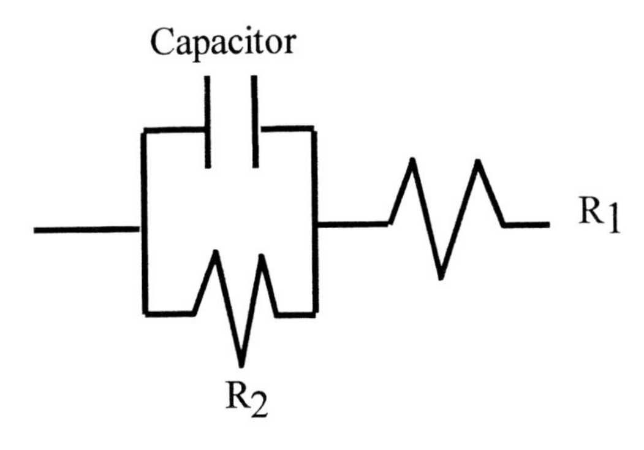

Figure 7.2. Electrical behavior of an EDL

EDL capacitive reactance is similar to the Figure 7.2 capacitor capacitance; R_2 is similar to corrosion resistance; and R_1 is similar to uncompensated solution resistance.

A capacitor takes time to reach full charge, and this charging-time produces a shift between current and voltage amplitude curves, as illustrated in Figure 7.3.[6] The shift is referred to as the phase angle and its magnitude is different for each polarizing voltage frequency. Phase angle is the difference between points on the X-axis where current and voltage curve amplitudes are zero, as shown in Figure 7.3. Phase angle magnitudes are often plotted as positive quantities for EIS data, even though their values are negative.

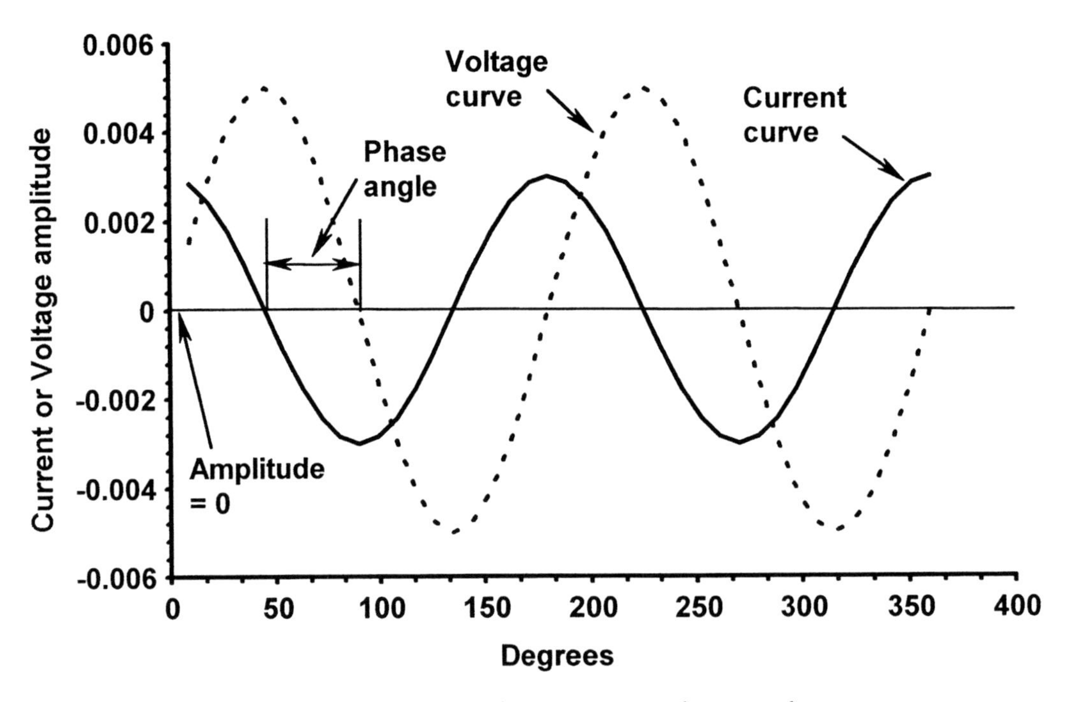

Figure 7.3. AC voltage-current phase angle

A resistor does not accumulate and separate charge when polarized by an AC (or DC) voltage. Consequently, a resistor does not exhibit time behavior (or frequency dependency) during polarization and the phase angle is zero.

7.5 Vector nature of AC polarizing voltage

AC (and DC) current and voltage are vectors because they have magnitude and direction. Consequently, impedance is also a vector because it is AC voltage divided by current. An impedance vector can be resolved into component vectors as illustrated in Figure 7.4,[7] where total impedance is represented by a solid arrow, and component vectors are dashed arrows. Phase angle is the angle between the X-axis and impedance vector.

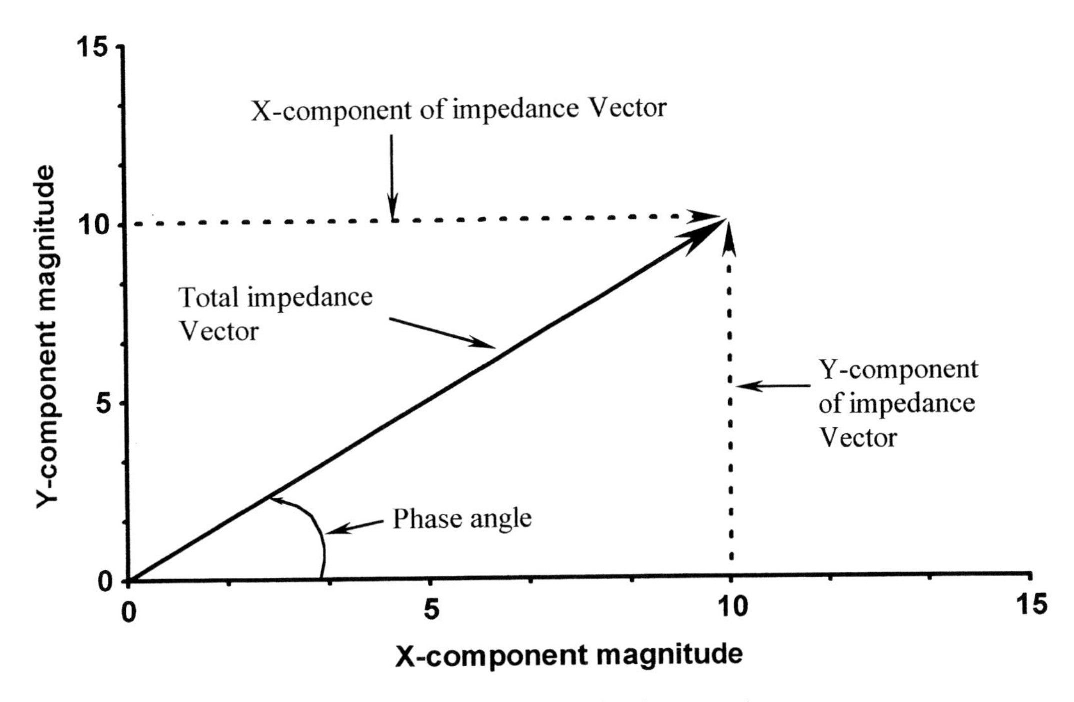

Figure 7.4. Vector nature of voltage and current

Total impedance magnitude for a circuit like that in Figure 7.2 is:[8]

$$Z = R_1 + \frac{R_2}{1 + (\omega R_2 C)^2} + \frac{j[-\omega R_2^2 C]}{1 + (\omega R_2 C)^2} \qquad [7.1]$$

where:

- Z is total impedance in ohms
- R_1 and R_2 are the resistors in Figure 7.2 (ohms)
- C is capacitor capacitance in Farads
- $\omega = 2\pi$(AC voltage frequency)
- j is the square root of -1 and is referred to as an imaginary number

Magnitude of the X-component vector in Figure 7.4 is equal to the second term in Equation 7.1, $R_2/[1 + (\omega R_2 C)^2]$, and is referred to as the real impedance. Magnitude of the Y-component vector is equal to the third term, $j[-\omega R_2^2 C]/[1 + (\omega R_2 C)^2]$, and is referred to as imaginary impedance because this term is multiplied by the imaginary number, j. Phase angle magnitude for a given frequency is the inverse tangent of imaginary impedance magnitude divided by real impedance magnitude. Equation 7.1 illustrates that each polarizing voltage frequency produces a different magnitude for phase angle, total impedance, and the component vectors.

Imaginary impedance magnitude is zero when the phase angle is zero, and the total impedance vector is on the X-axis. Conversely, the total impedance vector is parallel to the Y-axis (or on the Y-axis when there is no series resistor, such as R_1 in Figure 7.2) when phase angle in Figure 7.4 is 90 degrees.[9]

7.6 How an EDL responds to alternating current polarization

AC polarizations, like DC, also cause ions to move toward test and counter electrodes in an electrolyte, to maintain electrode electrical neutrality (see pages 20 and 21). However, ions move only in one direction for each DC voltage magnitude and polarity, but AC voltages cause ions to move back-and-forth between counter and test electrodes in response to changing voltage magnitude and polarity. Electron transfer also cycles to-and-from a test electrode and electrochemically active species during an AC polarization.

Figure 7.5 depicts a corroding metal EDL and the corresponding electrical circuit that describes: a) metal corrosion resistance, b) uncompensated solution resistance and c) EDL capacitive reactance.

Remember from Chapter 1 pages 5 through 9, that a test electrode EDL is an electrified interface that exhibits capacitive reactance to an applied voltage, and has resistance properties.[10] Consequently, an EDL has a time constant whose magnitude is equal to EDL capacitive reactance multiplied by the corresponding charge transfer (corrosion) resistance.

AC polarization forces an EDL like that in Figure 7.5 to try and change its chemical composition as fast as polarizing voltage frequency changes: a) magnitude and b) polarity (see pages 9 through 12). However, it takes time for EDL chemistry to change to a composition that produces, or corresponds to, a given polarizing voltage magnitude; much like it takes time for a capacitor to reach full charge after a voltage is applied.[11]

There is typically a range of frequencies, in an 100 killohertz to 5 millihertz EIS spectrum, where the time it takes to complete a polarization cycle is close to the time it takes for an EDL to reach steady state chemical composition (in response to a voltage magnitude and polarity change). That is, the time to reach EDL steady state chemistry is in-sink with polarization frequency. It is reasonable to conclude that an EDL time constant will be part of (and determine the location of) this range of frequencies, and that the response of an EDL to these frequencies will be different than the response to frequencies outside of this range. Section 7.9 will discuss how time constants responses are observed in EIS spectra.

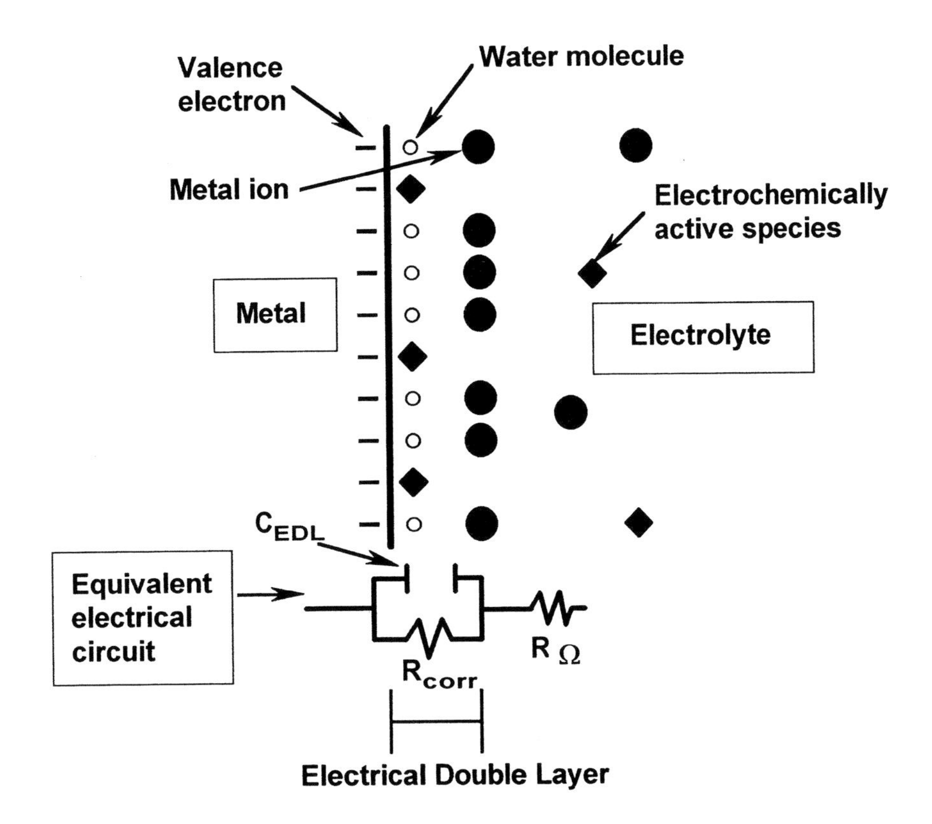

Figure 7.5. Electrical double layer for an uncoated, oxide-free corroding metal

Electrical double layers are not the only sources of test electrode time constants, and a test electrode can have more than one time constant.

7.7 Corroding coated metals: a two time constant system

Organic coating capacitive reactance and resistance properties can also produce a time constant. Water and ions typically diffuse into an organic coating, after a coated metal is submerged in an electrolyte, and change coating dielectric properties. Water and ions also move inside a coating in response to AC polarizations. However, water and ion movement through (and in) a coating is restricted by coating morphology, producing a coating, or pore resistance. Coating time constants can change with time, as electrolyte concentration in the coating changes.

Figure 7.6 depicts a corroding coated metal surface: a two time constant system. Small minus signs in Figure 7.6 represent negative ions (anions), positive signs represent positive ions (cations), and other symbols are the same ones used for Figure 7.5. The circuit for metallic corrosion in Figure 7.5 is nested inside the coating circuit. Nested circuits are used, instead of circuits in series, to indicate that pores in a coating can cause metallic corrosion by providing areas where the electrolyte has direct access to the metal surface.

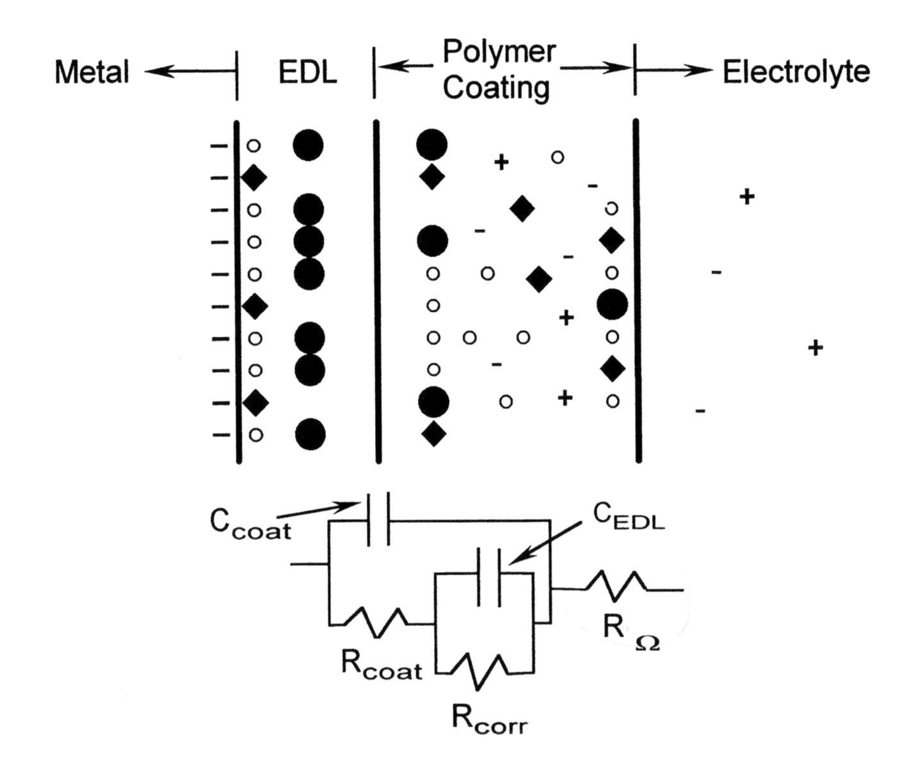

Figure 7.6. Electrified interface structure for a corroding, coated metal

7.8 How do polarization frequencies proceed through a two time constant system?

The number of time constants, and their magnitudes, determine how an AC polarization "moves through" a test electrode at any given frequency. Figures 7.7a through 7.7e use the coated metal electrical circuit diagram in Figure 7.6 to illustrate polarization pathways at different EIS polarization frequencies. This discussion is not technically rigorous, but it helps illustrate why different responses can be observed from the same test electrode and EIS spectrum. Discussion will proceed from highest to lowest frequencies.

High frequency AC voltages (e.g., 10^5 Hz) typically reverse magnitude and polarity so rapidly that EDL and coating capacitors do not impede coated metal polarization, and high frequency polarizations "pass-through" the coating capacitor as if it does not exist. Consequently, uncompensated solution resistance, R_{Ω}, is the only circuit component that responds (offers resistance) to high frequency polarizations, as depicted by the solid lines in Figure 7.7a below. Dashed vertical lines for the coating capacitor indicates that it does not significantly impede circuit (coated metal) polarization.

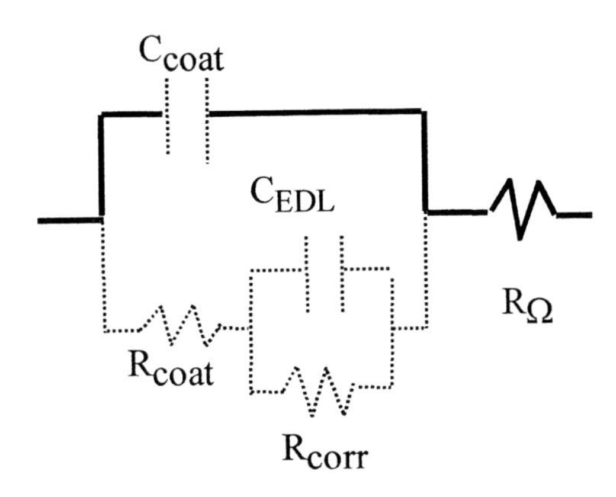

Figure 7.7a. Polarization pathway for high frequency polarizations

The polarization pathway when frequencies approach (the inverse magnitude of) the coating time constant is illustrated in Figure 7.7b. Coating time constants are typically smaller (higher frequency) than EDL time constants, so polarization frequencies in this range "pass through" the EDL capacitor as if it where not present in the circuit (coated metal). Consequently, the EDL capacitor is represented by dashed vertical lines.

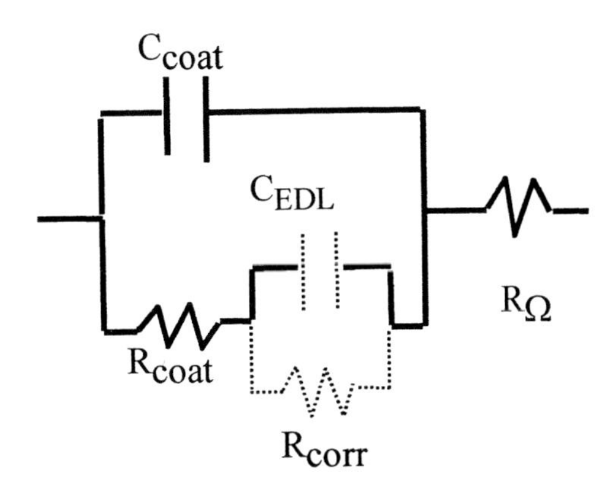

Figure 7.7b. Polarization pathway when frequency magnitude approaches the inverse magnitude of the coating time constant

Voltage frequencies between EDL and coating time constants are typically higher than (the inverse magnitude of) an EDL time constant, thus the EDL capacitor does not impede polarization of this frequency range. However, these frequencies are also low enough that they appear like a DC voltage to a coating capacitor. Consequently, a coating capacitor "blocks" polarization, and the pathway is that illustrated by solid lines in Figure 7.7c. Dashed vertical lines for the EDL capacitor indicates that it does not significantly impede polarization.

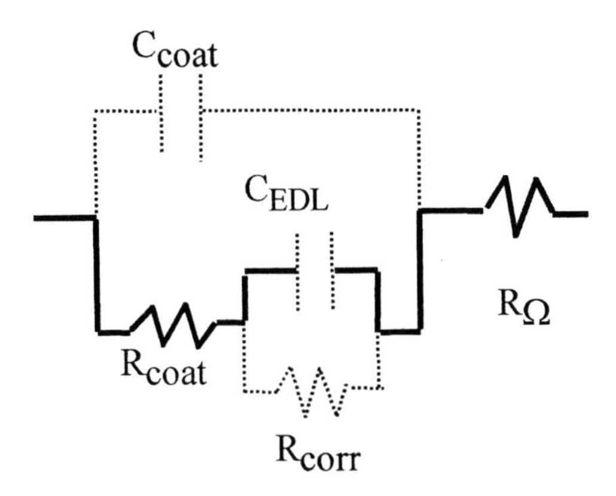

Figure 7.7c. Polarization pathway when polarization frequency magnitude is between the inverse magnitude for both time constants

Circuit polarization "proceeds through" the EDL capacitor and all the resistors when the polarizing frequency approaches (the inverse magnitude of) the EDL time constant, as illustrated by solid lines in Figure 7.7d. A portion of the polarization "leaks through" the capacitor so that impedance magnitude is less than what it would be if the pathway was only through the resistors.

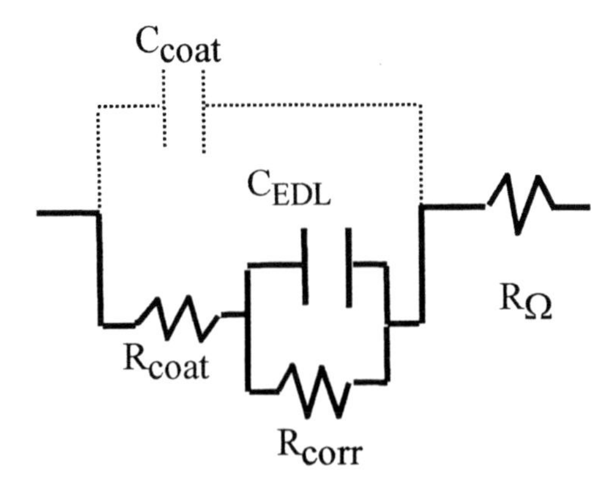

Figure 7.7d. Polarization pathway when frequency magnitude approaches the inverse magnitude of the EDL time constant

EDL and coating capacitances react to low frequency voltages (e.g., 0.005 Hz) as if they are DC voltages. Capacitors "block" DC voltage, so the low frequency polarization pathway is through all the resistors as illustrated by solid lines in Figure 7.7e.

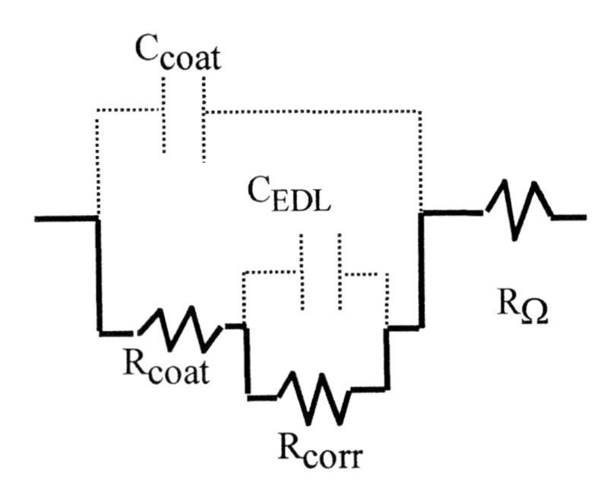

Figure 7.7e. Polarization pathway for low frequency polarizations

7.9 Common ways to graph EIS data

EIS data can be graphed a number of different ways using component vector magnitudes, total impedance magnitudes, or phase angles. Three of the most common types of EIS graphs are a) complex plane plots, b) Bode magnitude, and c) Bode phase plots.

A graph of real and imaginary impedance magnitudes (first plus second terms in Equation 7.1, and the third term in Equation 7.1, respectively) for each frequency is called a complex plane plot. Complex plane plots are also commonly called Nyquist plots.

Figure 7.8 contains a single time constant, complex plane plot, along with the analogous electrical circuit diagram. Imaginary impedance magnitudes are plotted as a function of real impedance magnitudes for each polarization frequency. The lowest polarization frequency is noted in parentheses on the right side of the semicircle, and the highest polarization frequency is noted in parentheses on the left side. The semicircle-shape in this type of plot is caused by the presence of the capacitor, or capacitive reactance. However, complex plane plots may look more like a straight line than a semicircle when the parallel resistor (e.g., corrosion resistance) is very large.

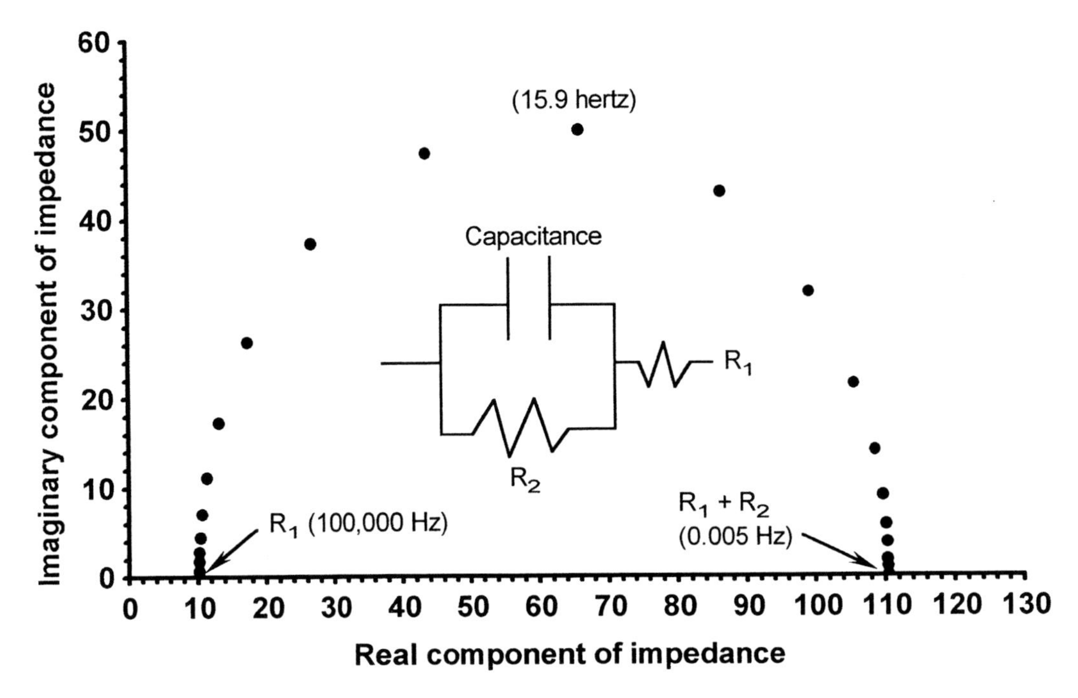

Figure 7.8. Single time constant complex plane plot

Uncompensated solution resistance corresponds to R_1 in the Figure 7.8, and corrosion resistance corresponds to R_2. The relationship between capacitive reactance magnitude, corrosion resistance, and the frequency at the semicircle apex is:[12]

$$C = \frac{1}{R_2\, \omega_{max}} \qquad [7.2]$$

Where "C" is capacitor capacitance, and ω_{max} equals 2π times the frequency at the complex plane plot semicircle apex.

Figure 7.9 contains the corresponding Bode magnitude plot for the Figure 7.8 complex plane plot. The X-axis of a magnitude plot contains log values of frequencies (base 10) and the Y-axis contains log values of total impedance for each frequency (Equation 7.1 on page 84). Locations for R_1, R_2, and capacitive reactance are also noted on the magnitude plot. Plot slope is zero when polarization is "through" resistances, and slope is less than zero (negative) when capacitive reactance becomes part of the circuit response to a polarization. Slope magnitude is determined by the ratio of R_2 to R_1, and approaches -1 as the ratio increases. For example, the slope is -0.27 when the ratio is 1:1, and -0.95 when the ratio is 1000:1.

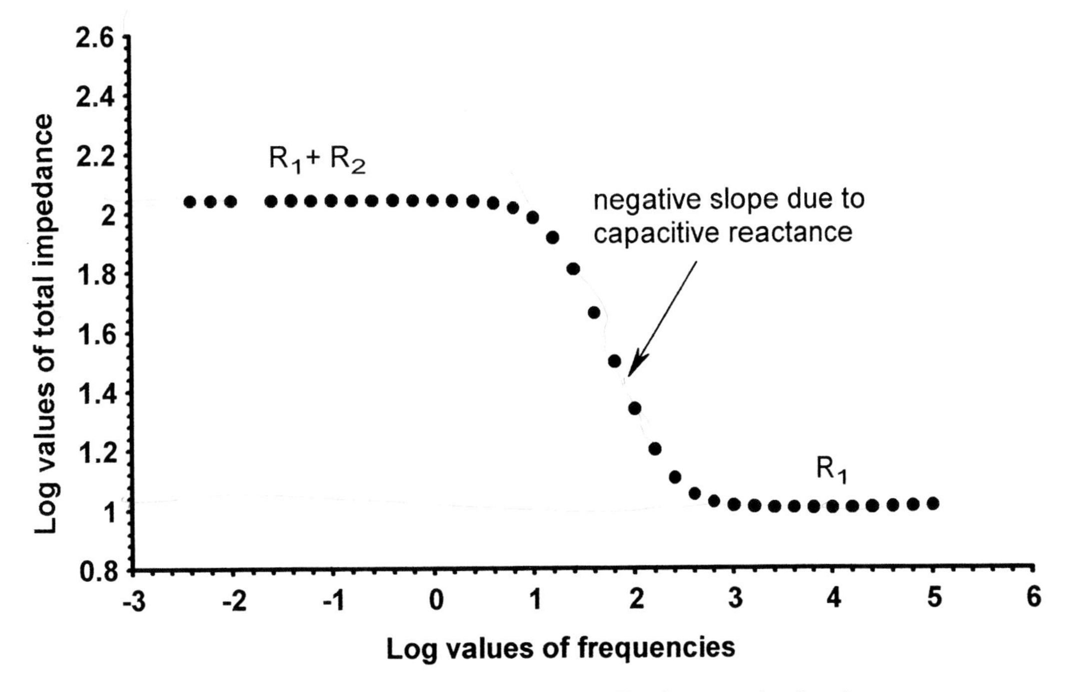

Figure 7.9. Single time constant Bode magnitude plot

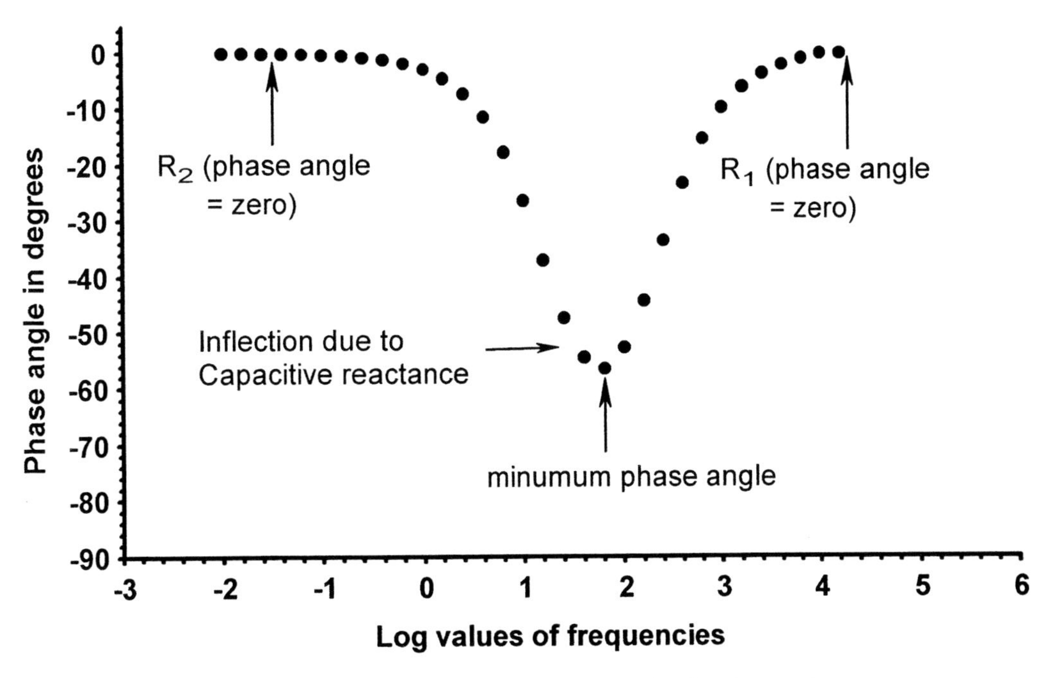

Figure 7.10. Single time constant Bode phase plot

Figure 7.10 contains the corresponding Bode phase plot for Figures 7.8 and 7.9. Resistances R_1 and R_2, and capacitive reactance (shown in Figures 7.8 and 7.9) are also noted in the phase plot. The inflection point frequency in Figure 7.10 is equal to:

$$\text{Inflection point frequency} = \frac{1}{1.77(R_2)C} \qquad [7.3]$$

Comparing Bode magnitude and phase plots shows that phase plot inflection corresponds to the area where the Bode magnitude slope is negative. Comparing phase and magnitude plots also illustrates that magnitude plot resistances, R_1 and R_2, correspond to polarizing voltage frequencies where phase angle is zero.

A variety of processes and surface films can also exhibit capacitive reactance and resistance properties, and produce responses in complex plane, Bode magnitude, and Bode phase plots. That is, each process or surface film can produce a distinct time constant.

7.10 Other sources of time constants

The magnitude of metal-to-electrolyte charge transfer (corrosion) resistance can be determined by metallurgical factors such as crystal defects and inclusions,[13] and type of electrolyte to which a metal is exposed.[14] It has also been demonstrated that different crystal orientations can corrode at different rates,[15-17] and it is well known that polycrystalline metal and alloy surfaces are not homogeneous.[18-20] Consequently, it is reasonable to hypothesize that polycrystalline metal surfaces have multiple charge transfer (corrosion) resistances and capacitive reactances that can produce multiple time constants.

However, multiple time constants produced by surface inhomogeneity or defects often have similar magnitudes, and are believed to cause complex plane semicircle or phase plot inflection point distortion, rather than producing distinct symmetrical semicircles, or sharp phase inflections for each time constant.[21] Surface roughness is also believed to produce multiple closely-spaced time constants that cause semicircle and inflection distortion.[22]

Restricting electrochemically active species diffusion to an electrode surface can also create different time constants.[23,24] Pitting corrosion has also been postulated as giving rise to data scatter at lower EIS frequencies.[25]

It is generally believed that time constants must be separated by at least one order of magnitude in order to avoid overlap in an EIS spectrum.[21] For example, coating capacitive reactance magnitudes are on the order of nanoFarads per square centimeter,[26-28] and EDL capacitive reactance magnitudes are typically on the order of microFarads per square centimeter.[29] A three order of magnitude difference between coating and EDL capacitive reactance magnitudes provides sufficient separation between time constants, and produces distinct complex plane semicircles for metallic corrosion and the coating.

The next Chapter contains an example of an EIS spectrum for a corroding coated metal, along with several examples of other types of EIS spectra.

7.11 Summary

This chapter discussed fundamental principles underlying electrochemical impedance spectroscopy (EIS) corrosion measurements. Capacitive reactance and resistance properties of an electrified interface produce time constants. Time constants produce a) inflections in Bode phase diagrams, b) semicircles in complex plane plots, and c) negative values for Bode magnitude plot slopes. There are a number of factors and processes that can produce multiple time constants on the same test electrode. A distinct response (e.g., phase plot inflection) can be observed for each time constant when time constant magnitudes are separated by at least one order of magnitude.

The ability of EIS spectra to measure separate processes occurring on the same electrode, is perhaps one of the most powerful features of EIS corrosion measurements.

7.12 References

1) R. Srinivasan, J. C. Murphy, C. B. Schroebel and R. S. Lillard, Materials Performance, pp. 14 - 18, March 1991
2) J. L. Dawson et. al., paper No. 125, Corrosion/78, Houston Tx, March 6 - 10, 1978
3) D. C. Silverman, Corrosion, **45** (10), pp. 824 - 830 (1989)
4) S. J. Downey and O. F. Devereux, Corrosion, **45** (8), pp. 675 - 684 (1989)
5) W. S. Tait, J. Coat. Technol., **61** (768), p. 57 (1989)
6) J. J. Brophy, Basic Electronics for Scientists, p. 102, McGraw-Hill Book Co., NY (1972)
7) I. Drooyan and W. Hadel, Trigonometry: an analytica approach, p. 210, Macmillan Company, NY (1967)
8) EG&G PAR technical application note AC-1
9) M. M. Sternheim and J. W. Kane, General Physics, p. 435, J. Wiley & Sons, NY (1986)
10) Impedance Spectroscopy: Emphasizing Solid Materials and Systems, J. R. MacDonald, editor, p. 2, J. Wiley and Sons, NY (1987)
11) Sternheim and Kane, Op. Cit., p. 346
12) R. Granata, E,G&G Princeton Applied Research application note AC-2
13) L. H. Van Vlack, Elements of Materials Science, second edition, p. 96, Addison-Wesley, Reading, MA (1964)

14) H. H. Uhlig, Corrosion and Corrosion Control, 2nd ed., p. 119, J. Wiley & Sons, NY (1971)

15) A. T. Gwathmey, Corrosion Handbook, edited by H. H. Uhlig, p. 33, John Wiley & Sons, NY (1948)

16) R. F. Mehl and E. L. McCandless, Trans. Am. Inst. Mining Met. Engrs., **143**, p. 246 (1941)

17) A. T. Gwathmey and A. F. Benton, Trans. Electrochem. Soc., **77**, p. 211 (1940)

18) H. S. Isaacs and M. W. Kendig, Corrosion **36** (6), p. 269 (1980)

19) L. J. Gainer and G. R. Wallwork, Corrosion, **35** (10), p. 435 (1979)

20) H. Leidheiser Jr., Corrosion, **39** (5), p. 189 (1983)

21) G. W. Walter, Corrosion Science, **26** (9), p. 681 (1979)

22) U. Rammelt and G. Reinhard, Corrosion Science, **27** (4), pp. 373 - 382 (1987)

23) M. Sluyters-Rehbach and J. H. Sluyters, Electroanalytical Chem.: A Series of Advances, **4**, pp. 16 - 17, A. J. Bard, editor, Marcel Dekker Inc., NY (1970)

24) R. D. Armstrong, M. F. Bell, and A. A. Metcalfe, Electrochemistry, **6**, pp. 98 - 127, Chemical Society Specialist Periodical Reports, UK (1978)

25) F. Mansfeld, Second International Symposium on Electrochemical Impedance Spectroscopy, July 12 - 17, 1992 in Santa Barbara, CA

26) N. Pebere, Th. Picaud, M. Duprat and F. Dabosi, Corrosion Science, **29** (9), p. 1073 (1989)

27) J. R. Scully, Report DTNSRDC/SME-86/006, David W. Taylor Naval Ship Research and Development Center, p. 105, Annapolis, MD (9186)

28) W. S. Tait and J. A. Maier, J. Coat. Technol., **62** (781), pp. 41 - 44 (1990)

29) A. J. Bard and L. R. Faulkner, Electrochemical Methods: Fundamentals and Applications, p. 8, John Wiley & Sons, NY (1980)

CHAPTER 8

Analyzing and Interpreting EIS spectra

Objectives:

After completing this Chapter, you will:

- know how to determine when EIS spectra contain multiple time constants
- be able to recognize induction in EIS spectra
- be able to recognize diffusion in EIS spectra
- be able to recognize when responses in EIS spectra are due to parasitic pathways
- know how to extract corrosion and coating parameters from single and multiple time constant EIS spectra
- know how to use capacitance to determine what type of process produces a given time constant
- know how to calculate corrosion rates from EIS spectra

8.1 Introduction

Fundamentals of EIS were discussed in Chapter 7, along with three examples of EIS graphs (plots) for a single time constant spectrum. This chapter augments Chapter 7 by providing examples of different single and multiple time constant EIS spectra. It is not within the scope of this book to provide examples of every possible type of spectra, so you will encounter spectra that are only somewhat similar to, or completely different from those contained in this chapter. However, there are enough different examples provided in this chapter to give you a critical mass of knowledge, so that you can interpret other types of spectra and be able to recognize when other spectra, in the corrosion literature, can also be applied to your systems.

8.2 Last time for terminology

New terms are defined in this section and throughout the chapter. Terms introduced in previous chapters will also be used, so you may want to review corrosion terminology definitions in previous chapters. Chapter 1 definitions are on pages 2 and 3; Chapter 2 on pages 18 and 19; Chapter 3 on pages 37 and 38; Chapter 4 on page 44; Chapter 5 on pages 53 and 54; Chapter 6 on page 64; and Chapter 7 on page 80.

Induction

The magnetic field produced by an AC current induces a voltage that resists making changes in current magnitude. The resistance-to-change produced by the magnetic field is referred to as induction.

Parameter

A parameter is a single number that characterizes system properties or behavior. Corrosion resistance and coating capacitance are examples of parameters that characterize corrosion behavior and coating performance, respectively.

8.3 EIS graphs for a corroding coated metal: a two time constant system

Chapter 7 Figures 7.8 through 7.10 (pages 89 through 91) contain single time constant complex plane, Bode magnitude, and phase plots, respectively. Figures 8.1 through 8.3 contain two time constant complex plane, Bode magnitude, and phase plots for epoxy coated type 1018 mild steel exposed to one molar potassium chloride electrolyte.[1] This system is a corroding coated metal like that depicted schematically in Figure 7.6 on page 86.

Notice that there are two semicircles in the Figure 8.1 complex plane plot; one for metallic corrosion at low frequencies and one for the coating at high frequencies.

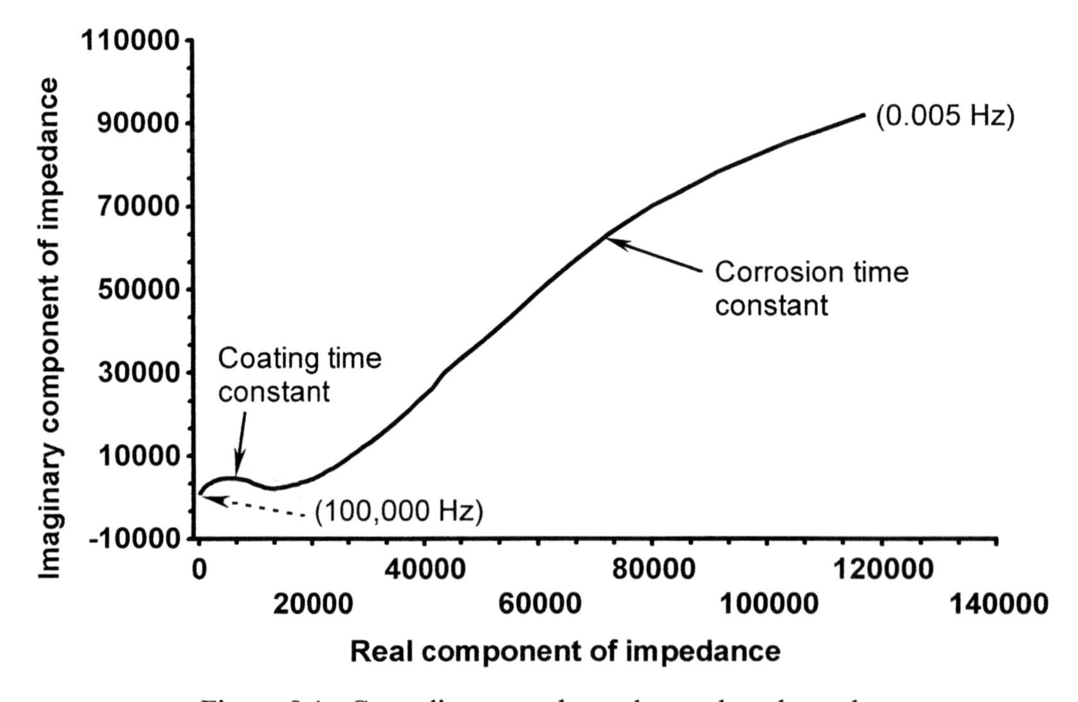

Figure 8.1. Corroding coated metal complex plane plot

The associated magnitude plot also has two time constants where slope approaches negative one in Figure 8.2; the associated phase plot in Figure 8.3 has two inflection points.

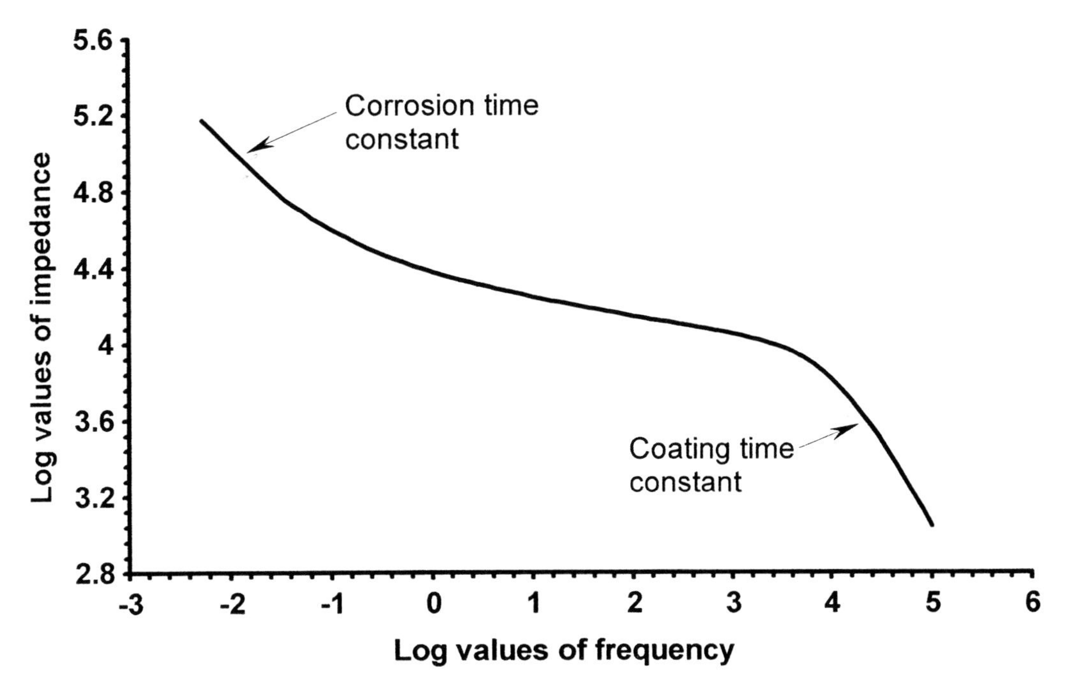

Figure 8.2. Corroding coated metal Bode magnitude plot

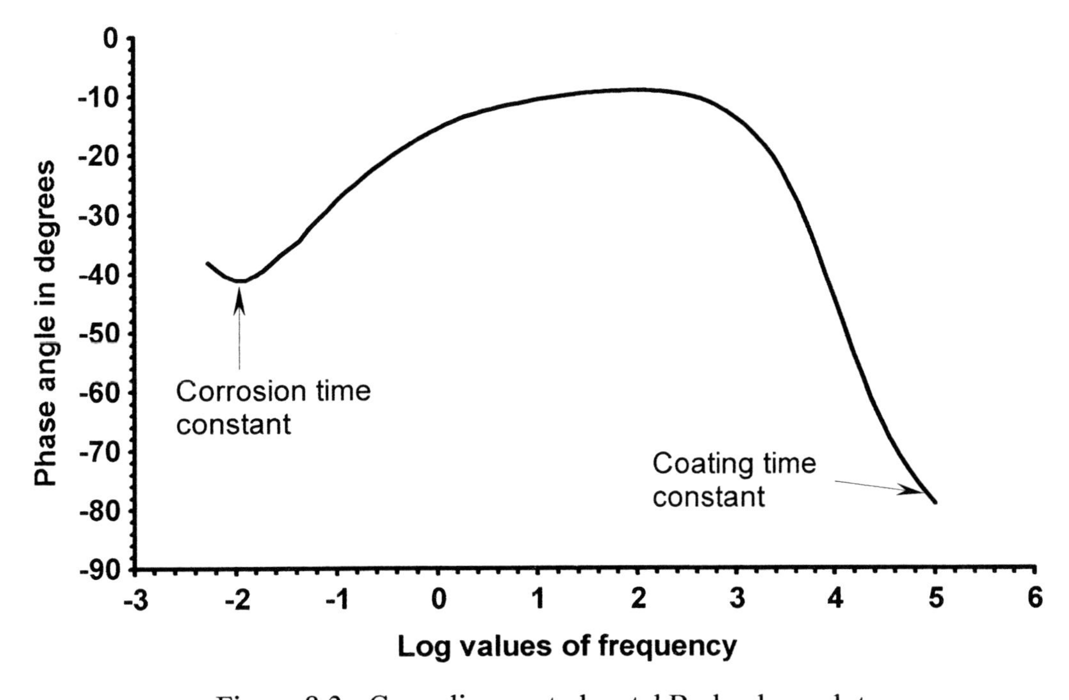

Figure 8.3. Corroding coated metal Bode phase plot

Sections 8.6 through 8.10 discuss how to extract parameters like coating and corrosion resistances, from multiple time constant EIS spectra, as well as how to determine what type of process is associated with a given time constant.

The next section contains complex plane, magnitude, and phase plots for a three time constant system, where it is not obvious from the complex plane or magnitude plots that there are three time constants.

8.4 Use all three types of EIS graphs as the first step in analyzing EIS data

There have been discussions on the merits of exclusively using one or another type of graph to analyze EIS spectra.[2] It is not an objective of this section to resolve this discussion, however, all three types of graphs can, and usually should be initially examined when analyzing EIS spectra. The next three figures illustrate why.

Consider the complex plane plot in Figure 8.4. It appears that there is only a single time constant. However, the associated magnitude plot in Figure 8.5 on the next page, appears to have several areas where curve slope approaches negative one (indicated by arrows). Examination of the associated phase plot in Figure 8.6 reveals that there are indeed three distinct time constants. Capacitance magnitudes for these time constants are consistent with those expected for a coated metal having no corrosion.[1]

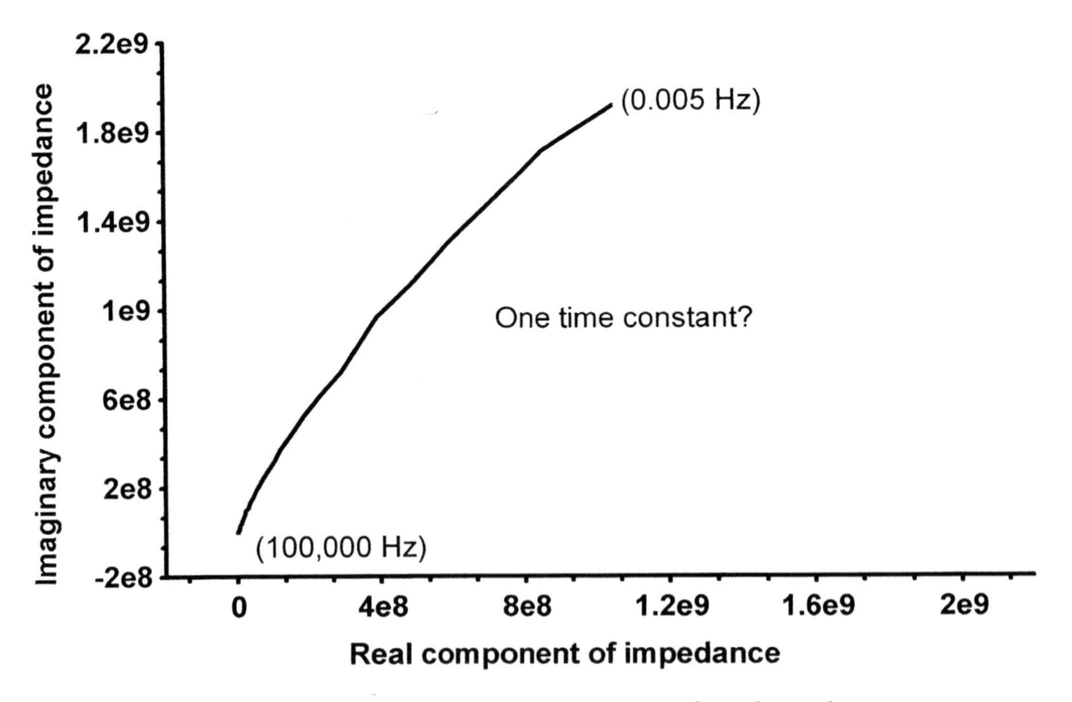

Figure 8.4. Triple time constant complex plane plot

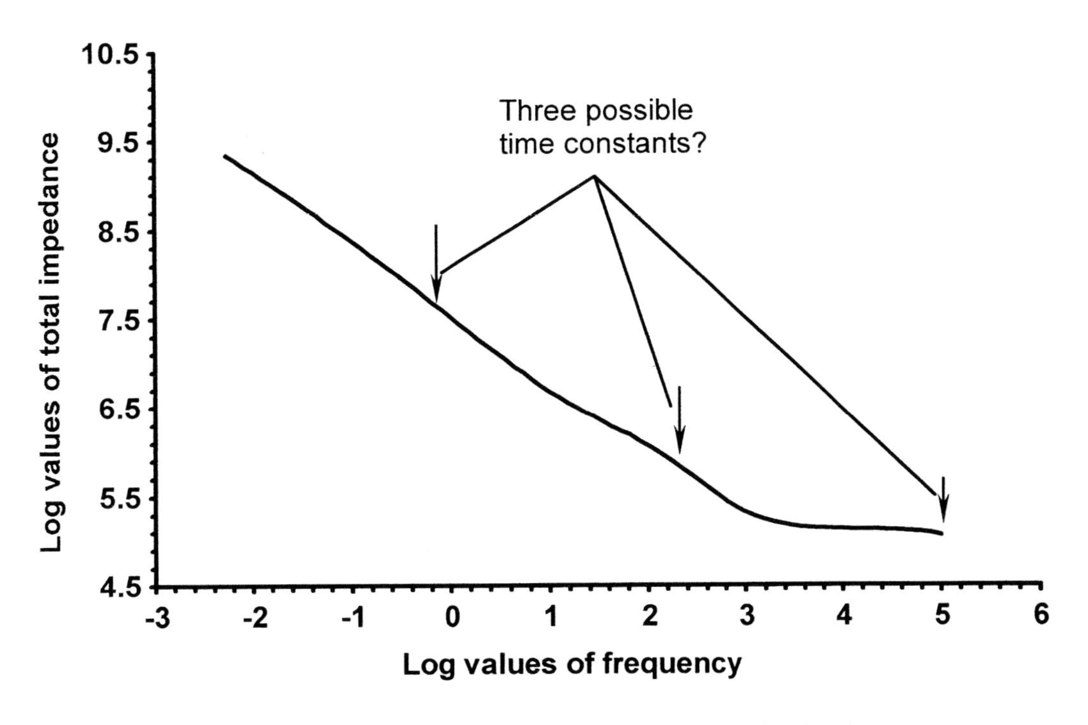

Figure 8.5. Triple time constant Bode magnitude plot

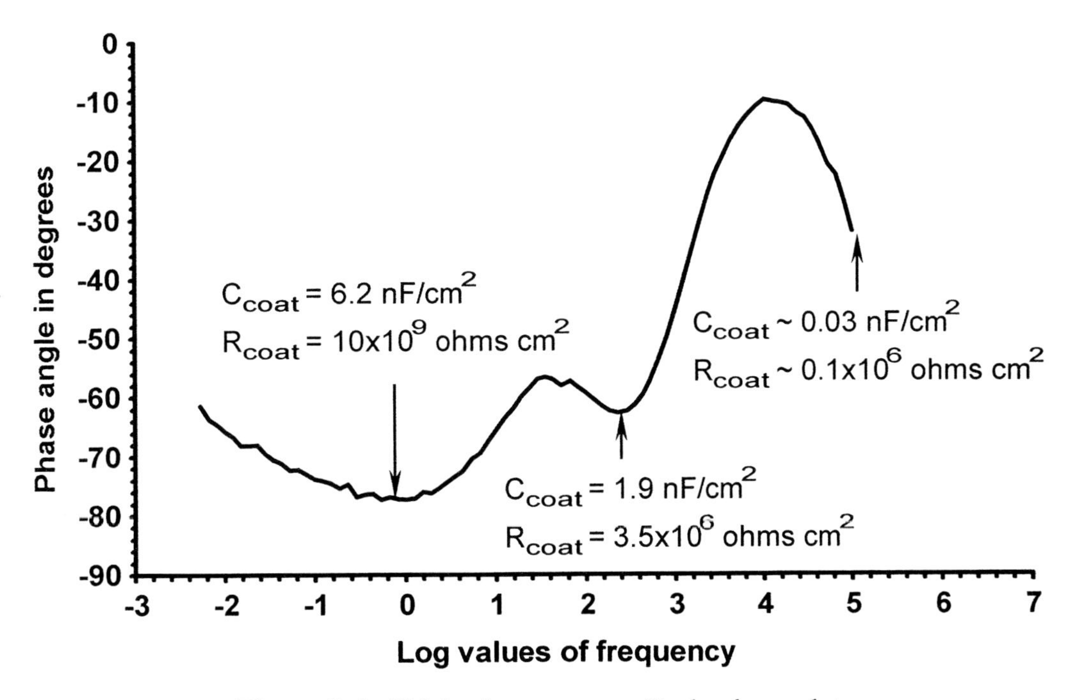

Figure 8.6. Triple time constant Bode phase plot

Figures 8.4 through 8.6 illustrate thc importance of examining all three types of EIS graphs. The next section contains examples of time constants produced by processes other than coatings and metallic corrosion.

8.5 Different types of responses in EIS spectra

Corrosion and coatings are not the only source of time constants, as discussed in Chapter 7 on page 92. This section of Chapter 8 contains examples of time constants produced by: a) induction, b) diffusion, and c) parasitic pathways.

A. Induction associated with a rapidly corroding uncoated metal

Electrochemical reactions often occur as a series of partial reactions, referred to as reaction steps. For example, the hydrogen reduction reaction is believed to proceed in the following series of reaction steps:[3]

1. hydrogen ions adsorb on a metal surface
2. hydrogen ions react with electrons to form hydrogen atoms
3. hydrogen atoms diffuse on the metal surface toward each other and combine to form hydrogen molecules
4. hydrogen molecules diffuse on the metal surface toward each other and combine to form hydrogen gas bubbles
5. hydrogen gas bubbles continue to combine with hydrogen molecules until the bubbles de-adsorb from the metal surface.

One or more of the individual steps in a complex reaction, like that for hydrogen reduction, can be significantly slower than the others and the slowest step can restrict how fast electrical double layer (EDL) chemical composition changes in response to polarizing voltage magnitude and polarity changes. Restricting EDL chemistry changes can produce induction. Adsorption or desorption equilibrium of surface active species, or corrosion inhibitors, on a metal surface have also been reported as a possible source of induction in EIS spectra.[4]

Figure 8.7 contains a single time constant complex plane plot having induction, obtained from uncoated type 1018 mild steel in hydrochloric acid. Induction produces negative impedance magnitudes at high and low frequencies, as seen in Figure 8.7.

Figures 8.8 and 8.9 contain the associated Bode magnitude and phase plots. Induction also produces negative impedance magnitudes in a magnitude plot, and positive phase angles in a phase plot, as seen in Figures 8.8 and 8.9.

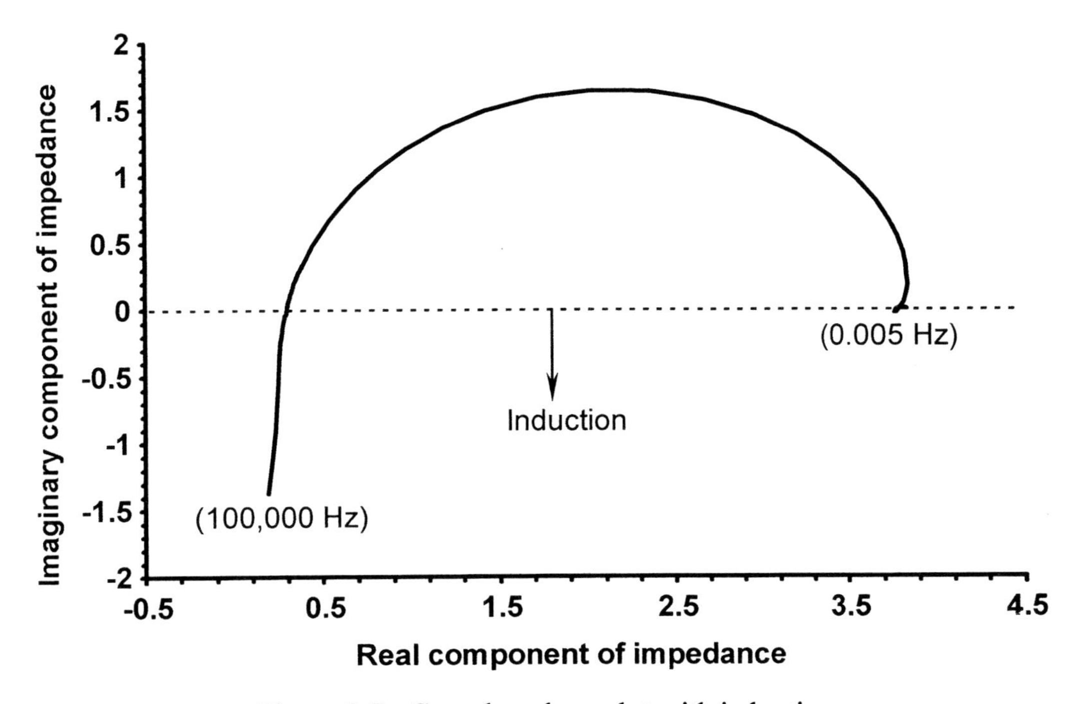

Figure 8.7. Complex plane plot with induction

Induction also distorts complex plane semicircles, but corrosion resistance is usually easy to obtained from the associated Bode magnitude plot at low frequencies.

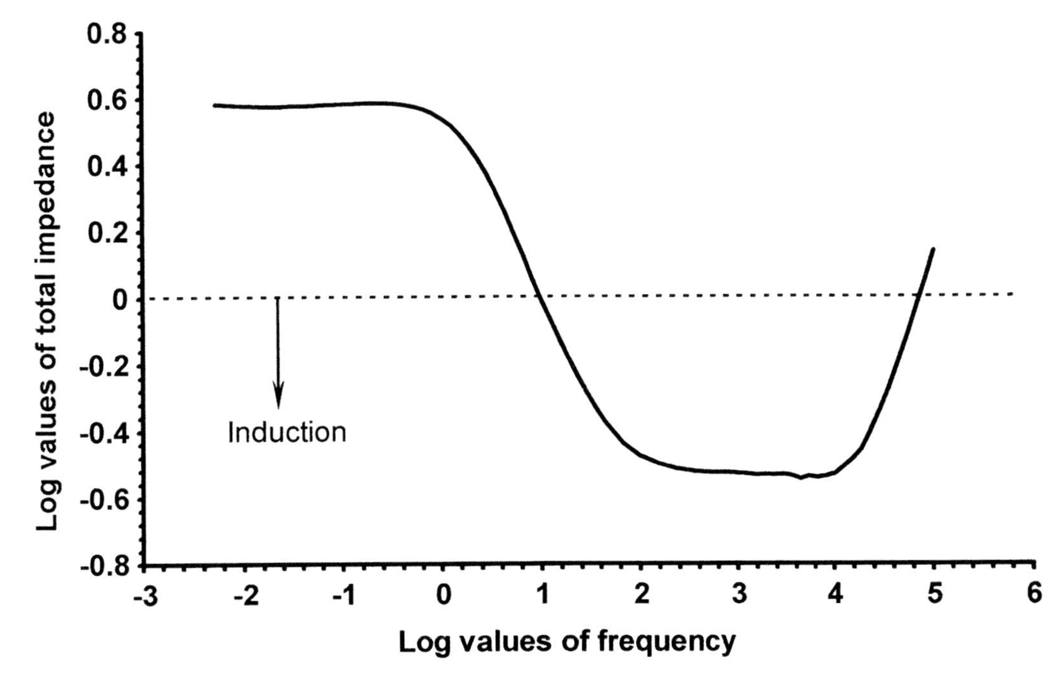

Figure 8.8. Bode magnitude plot with induction

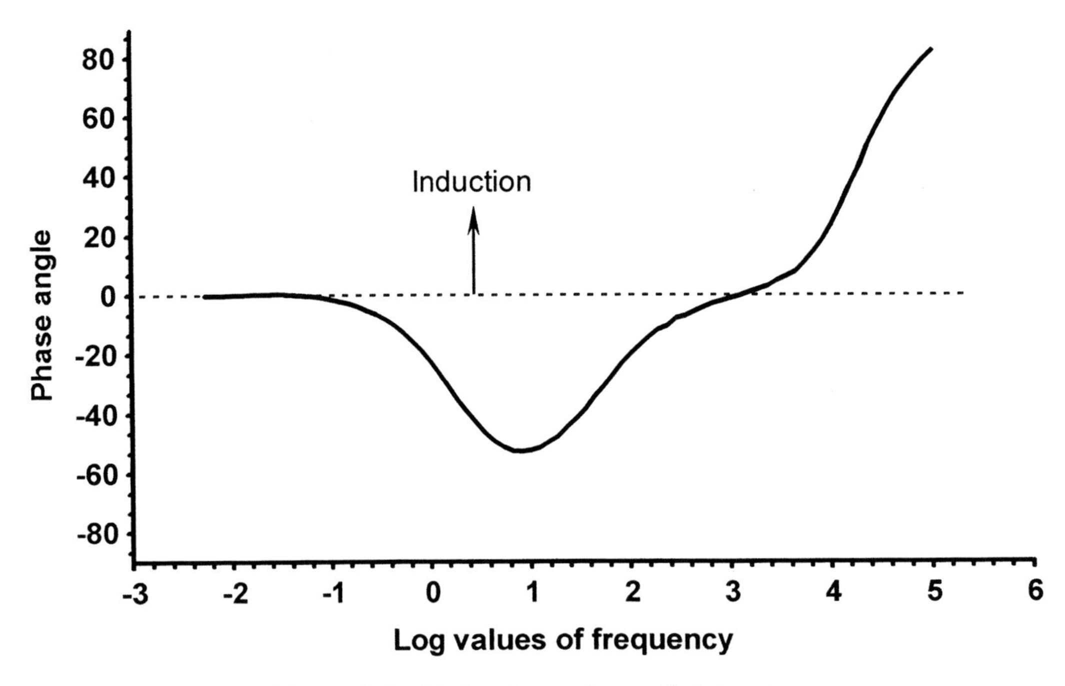

Figure 8.9. Bode phase plot with induction

B. Responses produced by diffusion

Sometimes electron exchange between a metal and an electrochemically active species is so rapid that the corrosion rate is controlled by the rate of electrochemically active species diffusion to a metal surface, or the rate at which corrosion products desorb and diffuse away from a metal surface. For example, corrosion products like hydrogen gas bubbles can block access of electrochemically active species to a test electrode surface, thereby restricting hydrogen ion diffusion to the surface and thus impede the overall corrosion reaction rate. EIS spectra can contain time constants for both corrosion and diffusion when diffusion controls or limits a corrosion rate. Diffusion time constants are, typically, observed at lower frequencies; corrosion time constants are, typically, observed at higher frequencies.

Figures 8.10 through 8.12 contain EIS graphs for situations where the electrochemically active species diffusion path to an electrode is considered to have an infinite length and is referred to as a Warburg diffusion.[5] Areas of the spectrum associated with corrosion and diffusion are noted in the Figures 8.10 through 8.12.

Notice that the low frequency end of the Figure 8.10 complex plane plot is approximately a straight line, instead of a large diameter semicircle like that for the corroding coated metal complex plane plot in Figure 8.1.

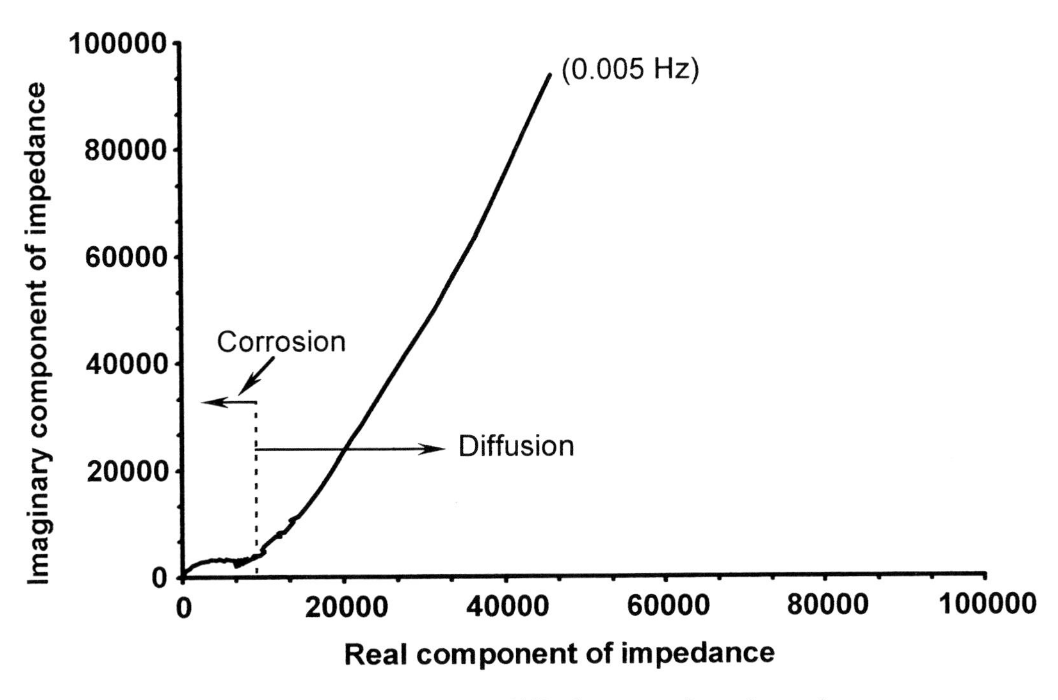

Figure 8.10. Warburg diffusion complex plane plot

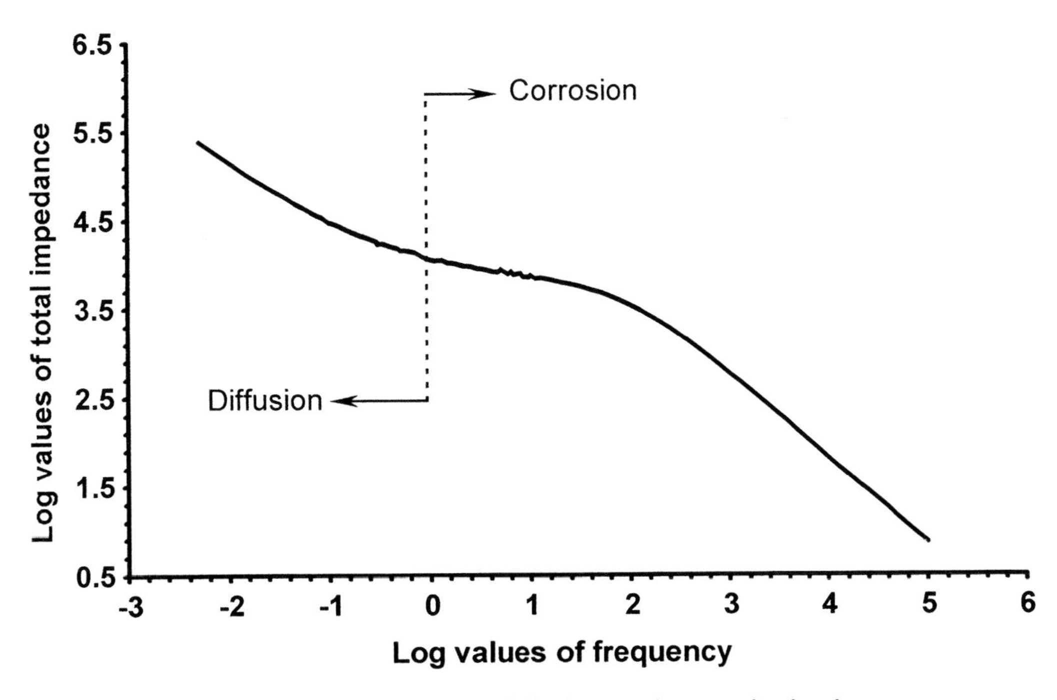

Figure 8.11. Warburg diffusion Bode magnitude plot

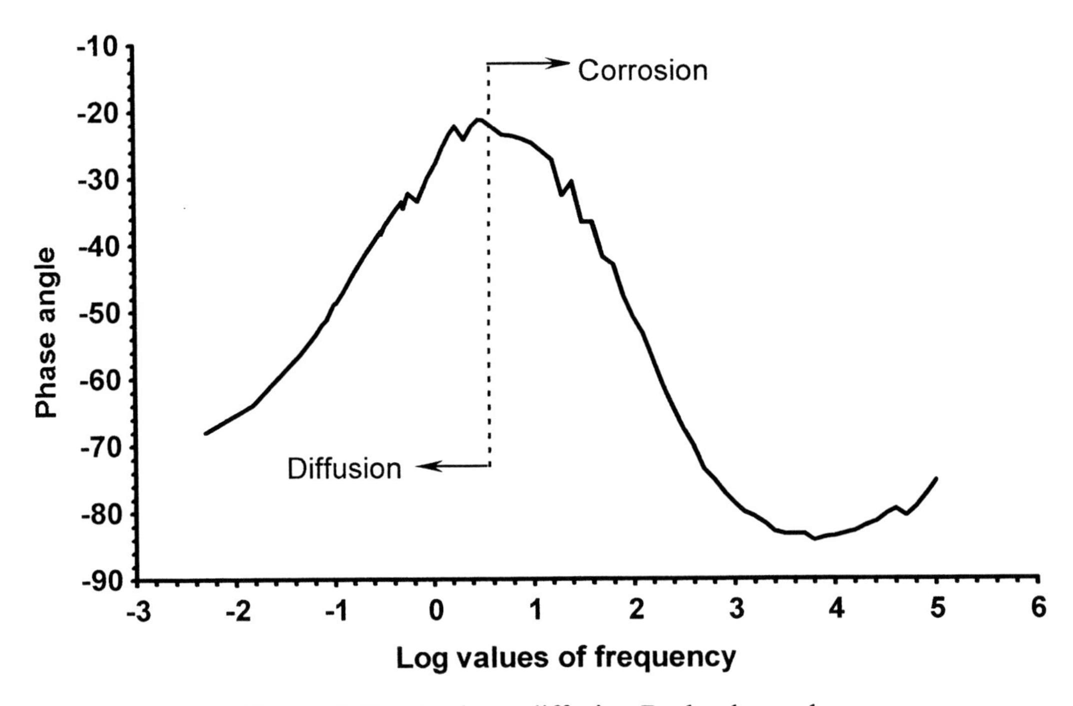

Figure 8.12. Warburg diffusion Bode phase plot

Figures 8.13 through 8.15 contain EIS graphs in which the diffusion path for electrochemically active species is considered to have a finite length.[6] The low frequency complex plane time constant in Figure 8.13 appears to be a "flattened-out" semicircle, instead of a straight line like that for Warburg diffusion in Figure 8.10.

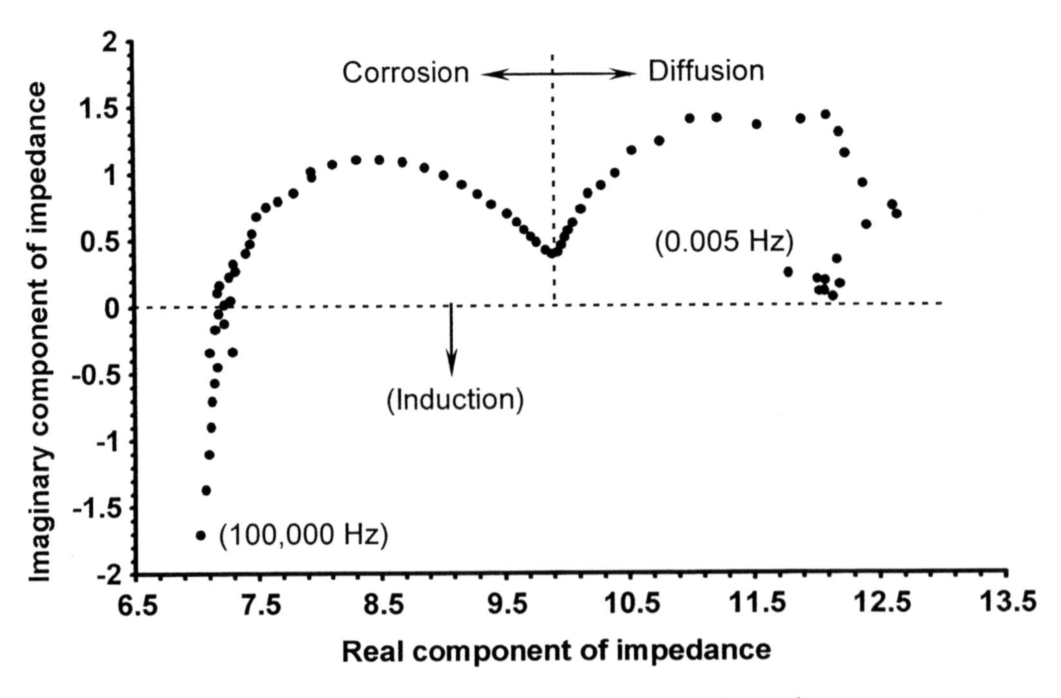

Figure 8.13. Finite diffusion complex plane plot

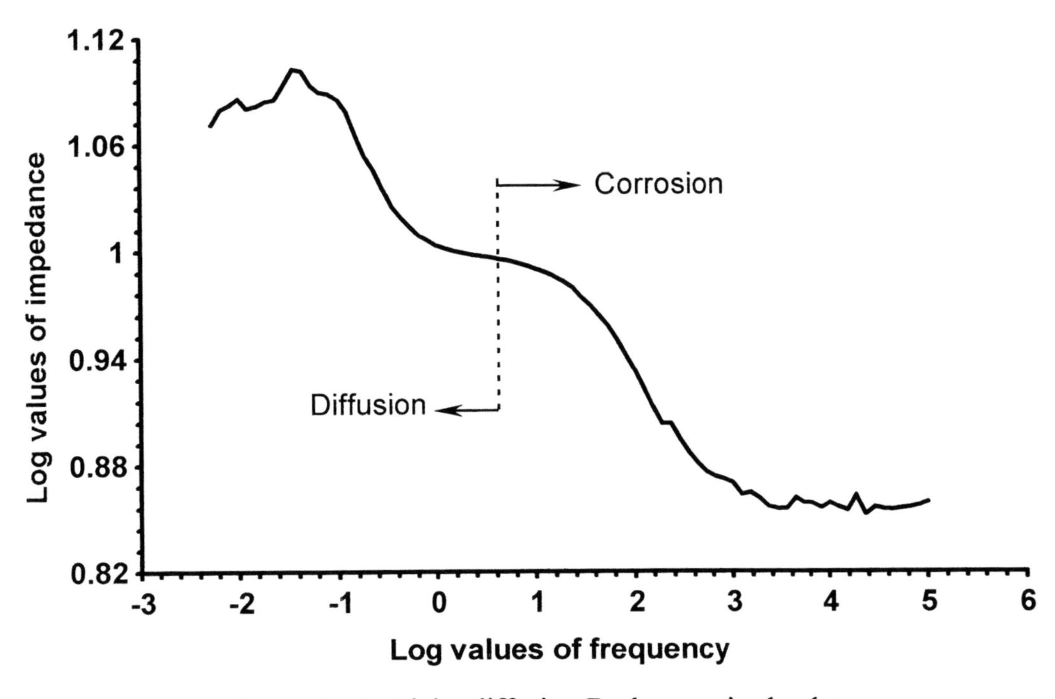

Figure 8.14. Finite diffusion Bode magnitude plot

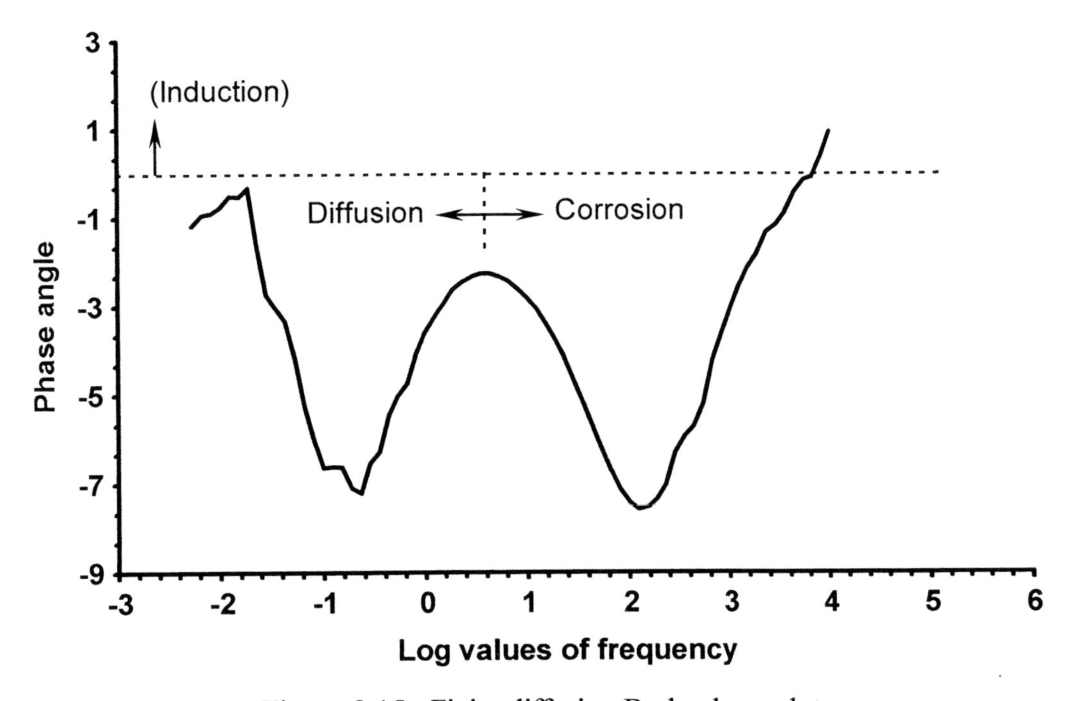

Figure 8.15. Finite diffusion Bode phase plot

It can be verified that a low frequency time constant is produced by Warburg, or finite length diffusion, by increasing electrolyte stirring and observing if the resistance for a low frequency time constant decreases or, in some cases, disappears.[6] A low frequency time constant is most likely not due to diffusion if resistance does not decrease (or disappear) with increased stirring.

C. Responses produced by corrosion measurement equipment

There are situations where electrolyte solution resistance is so high (e.g., hydrocarbon solvents containing trace water) that applied AC voltage finds a less resistive path through test equipment electronic circuitry, instead of through the test electrode.[7,8] Such a pathway is referred to as an equipment artifact or parasitic pathway, that produces time constants that look entirely plausible.

A spectrum containing time constants for corrosion and a parasitic pathway is contained in the complex plane plot in Figure 8.16. This spectrum was obtained from uncoated 304 stainless steel in 18 Mohm-cm resistivity deionized water. The low frequency time constant looks like a Warburg because only a portion of the real and imaginary data were used, so that the high frequency time constant could be easily observed. The low frequency time constant is actually a large diameter semicircle.

Figures 8.17 and 8.18, on the next page, contain the associated magnitude and phase plots.

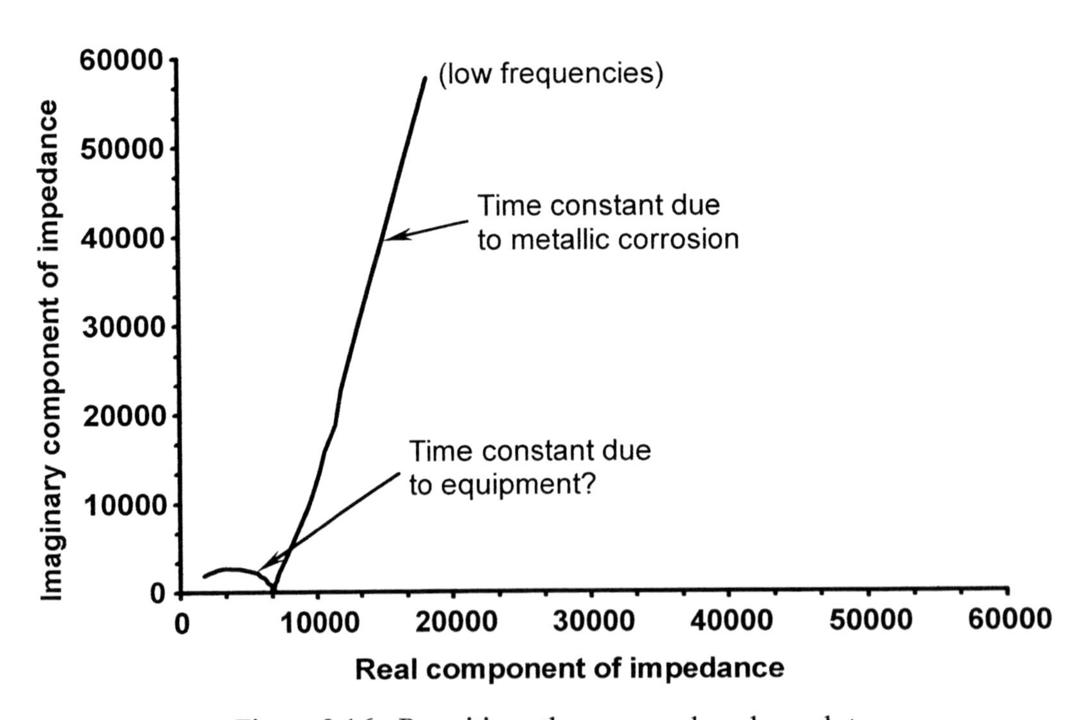

Figure 8.16. Parasitic pathway complex plane plot

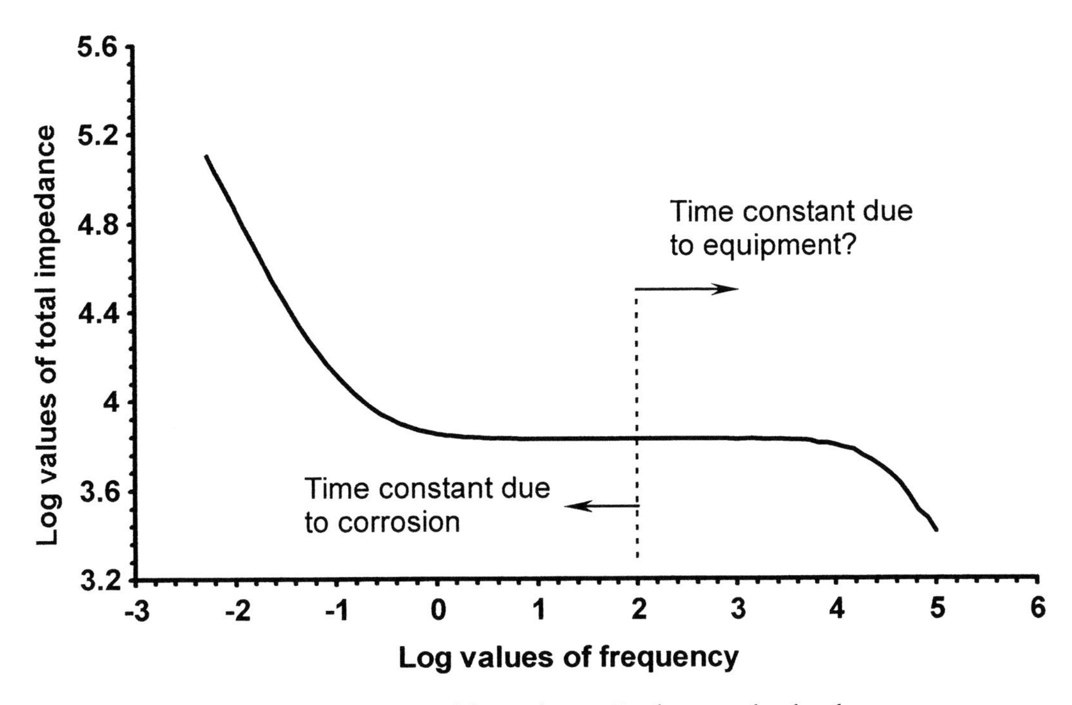

Figure 8.17. Parasitic pathway Bode magnitude plot

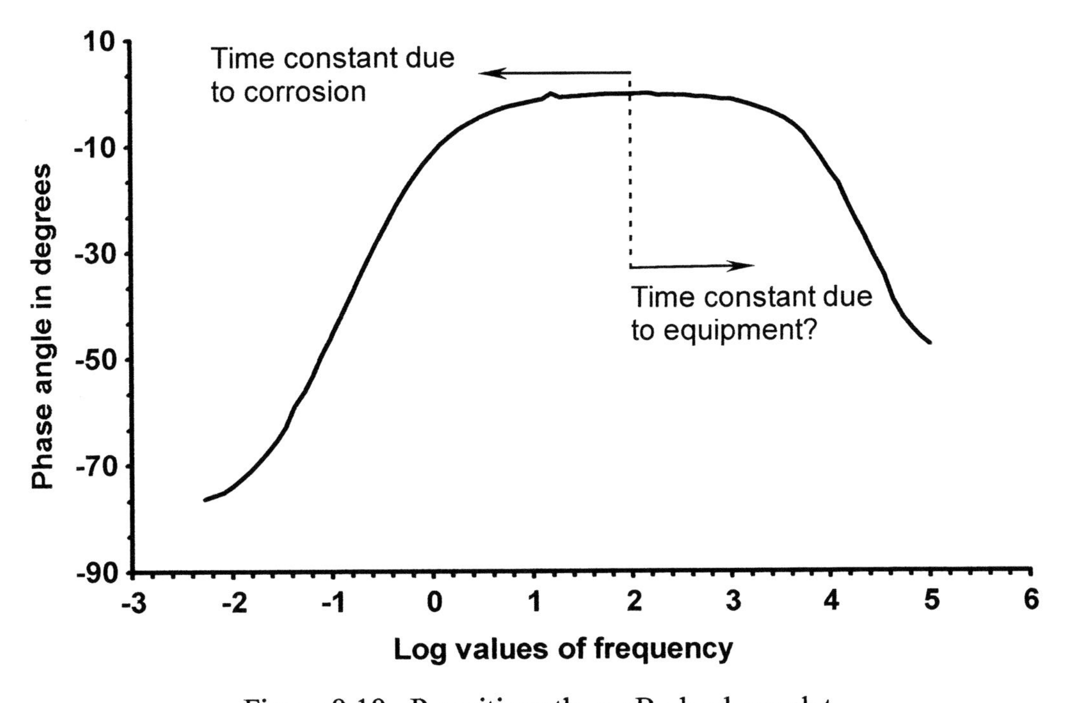

Figure 8.18. Parasitic pathway Bode phase plot

Uncompensated solution resistance is 3.12×10^3 ohm-cm^2, for Figures 8.16 through 8.18, which is the same order of magnitude as the resistance for the high frequency time constant. Consequently, it is reasonable to conclude that the high frequency time constant in Figures 8.16 through 8.18 is due to a parasitic pathway and not the stainless steel electrode.

Reference number 7 contains a detailed discussion on how to remove parasitic pathway time constants from EIS spectra.

8.6 Extracting corrosion and coating parameters from EIS data

Parameters obtained from EIS spectra are used to: a) characterize corrosion and coating behavior, and b) estimate or predict equipment service lifetime. Parameters can be extracted either graphically, or through use of computer software.

A Bode magnitude plot can be used to graphically estimate capacitance magnitudes by extrapolating each capacitance portion of the plot (areas where slope approaches negative one) to impedance values where $\log(2\pi f) = 0$ (approximately -0.798 on the frequency axis). Time constant capacitive reactances are equal to inverse impedance when $\log(2\pi f)$ is zero on a magnitude plot. Time constant resistances are equal to impedance where magnitude plot slope is zero. Phase angle maxima can be used to help locate areas where magnitude plot slope is zero, when it is difficult to find these areas on the magnitude plot.

There are several computer software programs for extracting capacitance and resistance magnitudes from EIS spectra, making it unnecessary to use graphical analysis. A partial listing of these programs is:

- CIRFIT developed by Kendig and Mansfeld[9]
- EQUIVALENT CIRCUIT developed by Boukamp[10]
- LEVM developed by Macdonald[11]
- Model 398 Impedance measurement software developed by EG&G Princeton Applied Research
- ZFIT developed by Walters[12]
- Zplot and Zview, distributed by Solartron Instruments and Scribner and Associates

EG&G Princeton Applied Research and Solartron software packages are also used to collect and graph EIS data.

It is not within the scope of this book to discuss merits and/or shortcomings of software programs and each software package manual contains instructions on how to use software to extract parameters from EIS spectra.

8.7 Using equivalent electrical circuit models to obtain parameters from EIS spectra

The most technically rigorous way to obtain parameters that characterize corrosion or coating behavior, is to build a mathematical equation (i.e., model) from test electrode corrosion reaction kinetics and regress (fit) this equation around experimental EIS data.[6] Unfortunately, corrosion kinetics for industrial systems are often very complex and/or unknown and an exact mathematical model can not be developed for regressing around the data.

Fortunately, EIS parameters can be estimated by using mathematical equations for simple electrical circuits. Figure 8.19 contains an example of such an electrical circuit and its mathematical equation, that can be used to model symmetrical, single time constant EIS spectra. This circuit and equation were discussed in Chapter 7 on pages 85 and 83, respectively.

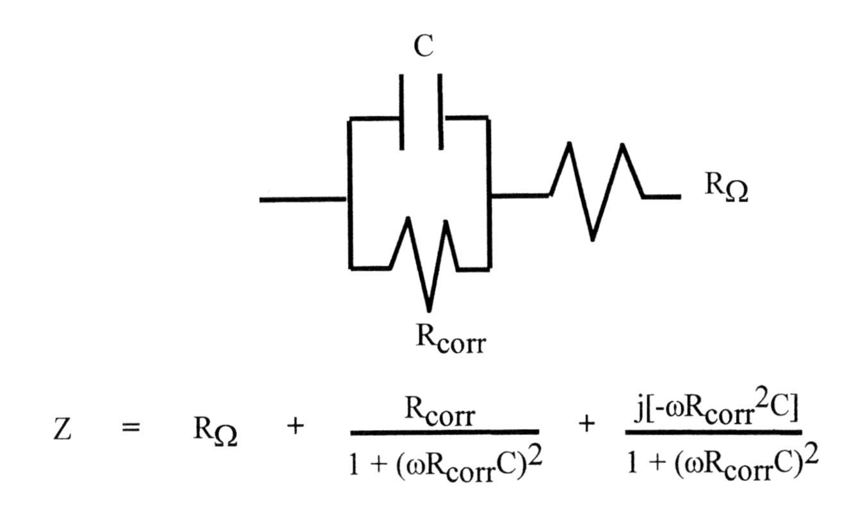

$$Z = R_\Omega + \frac{R_{corr}}{1 + (\omega R_{corr} C)^2} + \frac{j[-\omega R_{corr}^2 C]}{1 + (\omega R_{corr} C)^2}$$

Figure 8.19. Simple electrical circuit and its associated mathematical equation

Software packages mentioned in section 8.6 regress an equation (like that in Figure 8.19) around experimental data, by iteratively adjusting parameters, such as R_{corr} and C in Figure 8.19, until the closest fit of an equation to experimental data is obtained. Statistical tests are used to determine when equation parameters have been adjusted to give the closest equation fit to the data. More will be said in the next section about using statistical tests to help fit equations to data and decide when the best equation for the data has been obtained.

A circuit equation can be fit to an entire spectrum, as illustrated in Figure 8.20, or individual time constants can be modeled separately using a modified Randles circuit, as illustrated in Figure 8.21.[1]

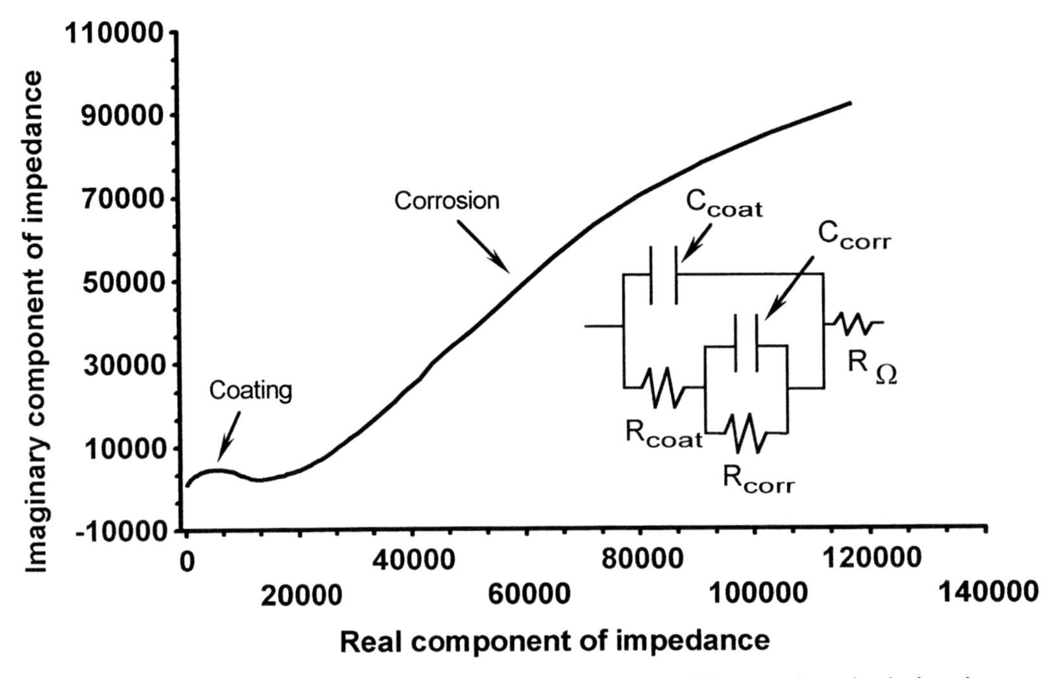

Figure 8.20. Modeling an entire EIS spectrum with one electrical circuit

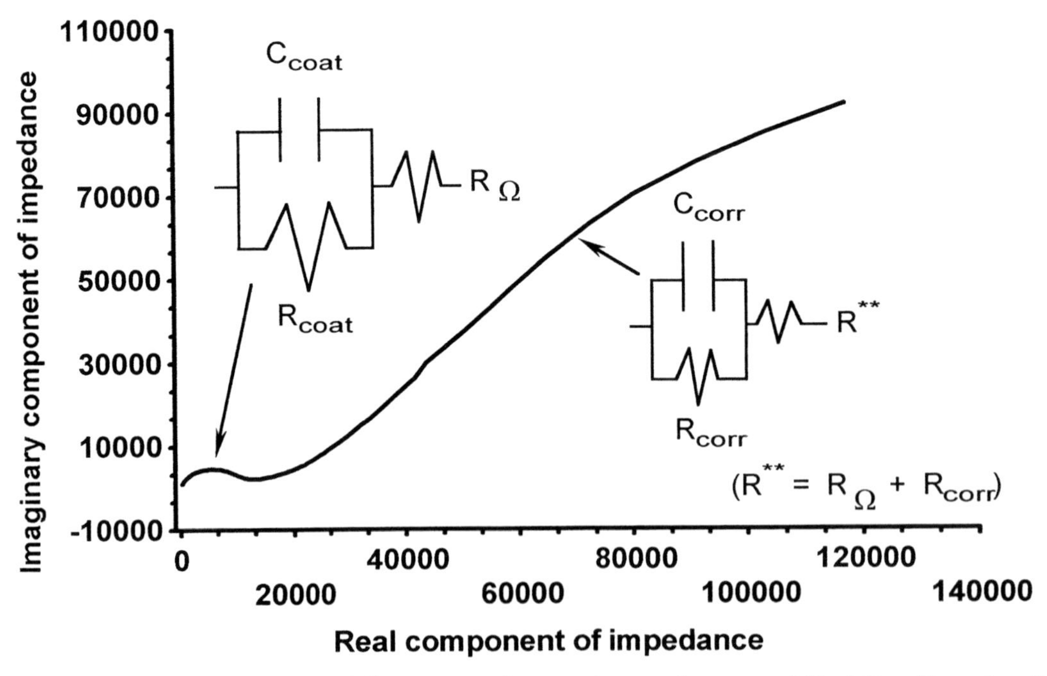

Figure 8.21. Modeling an EIS spectrum by sections using a modified Randles circuit

Parameters obtained from separately modeling time constants may not be as accurate as those obtained from modeling the entire spectrum. However, estimated parameters obtained from separately modeling time constants are often accurate enough for industrial corrosion and service lifetime estimations.[1]

8.8 How do I know that I have the best estimates for coating and corrosion parameters?

Resistance and capacitance magnitudes for equations like that in Figure 8.19, are iteratively adjusted until differences between experimental data and corresponding calculated values are at their smallest possible magnitudes for each frequency. These differences are referred to as residuals[13]; the process of adjusting parameters to obtain the smallest residual magnitudes is referred to as: regression or modeling.[14] An ideal regression (model) has all residuals equal to zero, but the summation of residuals must have the smallest possible magnitude to obtain the best fit of a given model to experimental data.

Correlation coefficients are often reported with models, but correlation coefficients only measure strength of the relationship between dependent and independent variables and not causation.[15] Consequently, a correlation coefficient is not a measure of how good a model fits experimental data and thus how close model parameters are to actual parameter magnitudes. Hence, a correlation coefficient should not be used by itself to validate a model.

Statistical tests should be applied to each regression to evaluate how well an equation (model) fits the experimental data. Not testing model validity can produce erroneous estimations for corrosion and coating resistance parameters. There are several statistical tests that can be used to assess goodness of model fit to experimental data. Two of these tests are the Chi square statistic, χ^2, and analysis of residuals.

The χ^2 statistic is calculated from the sum of the squares of all residuals,[16,17] and is a direct measure of how close a model fits experimental data. An ideal model has a χ^2 of zero, however, actual χ^2 values less than 1 are considered sufficient to indicate a good fit of a model to experimental data.

Minimizing residual magnitudes is the key to obtaining a good model fit to experimental data. There are two important assumptions associated with the regression process: 1) residual values are randomly distributed among all frequencies, and 2) the sum of all residuals is zero, or nearly zero.[13,18] Consequently, a plot of residuals as a function of polarizing voltage frequency, referred to as a residuals plot,[19] provides a statistical method for determining how well a model fits experimental data. Figure 8.22 contains a residuals plot, where total impedance residuals are plotted as a function of polarizing voltage frequency.

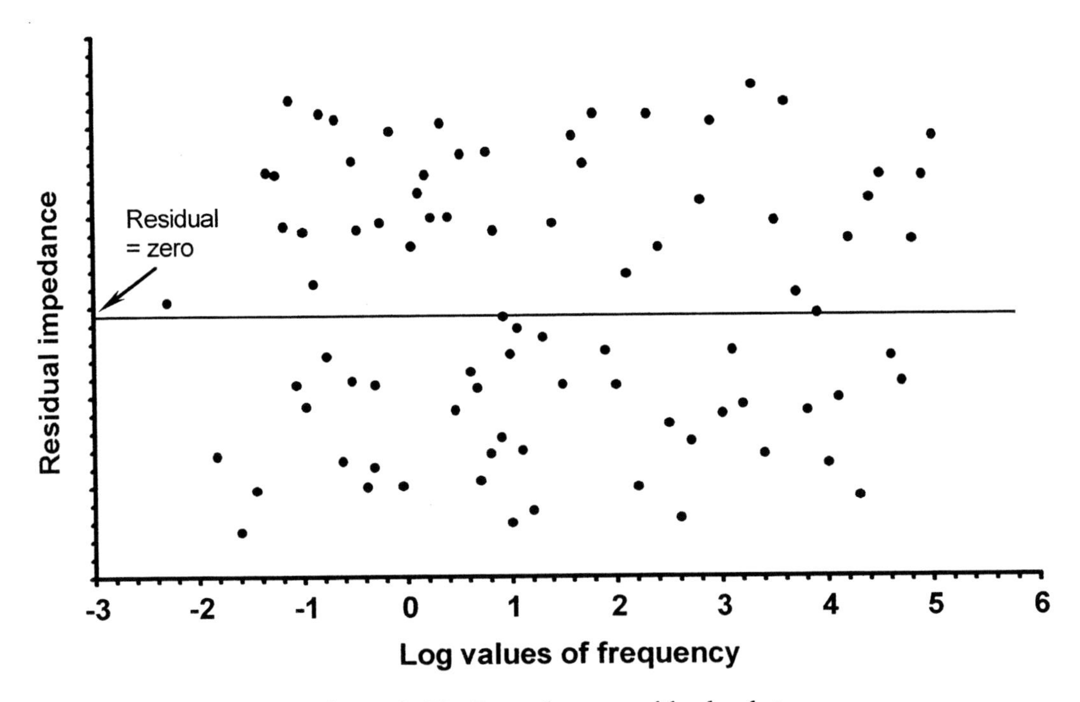

Figure 8.22. Impedance residuals plot

A model is a good fit to experimental data when the number of residual values $\geq$ zero are equal to the number of values $\leq$ zero, as illustrated in Figure 8.22. Model validity should be questioned when significantly more residual values are above or below zero, or periodicity of residuals is observed around zero.[13]

Some of the software packages listed in section 8.6 give users the option to apply weighting factors to individual data points in order to improve model fit. This approach is valid as long as there are sufficient degrees of freedom in the data.

Each polarizing voltage frequency has one degree of freedom, and each mathematical operation on a spectrum, such as a weighting factor or parameter, is another degree of freedom. For example, the equation in Figure 8.19 has three adjustable parameters, R_{Ω}, R_{corr}, and C, each having one degree of freedom. The process of doing a regression on a spectrum consumes one degree of freedom, thus a total of four degrees of freedom are needed to regress the equation in Figure 8.19 around an EIS spectrum. Consequently, more than four frequencies are needed to validly fit a simple circuit like that in Figure 8.19, to an EIS spectrum, and the largest number of weighting factors permissible are the number of frequencies in the data minus four degrees of freedom.

There are occasions where a model can not be found, or weighted, that does not avoid residual periodicity, or non-random distribution of residuals above or below zero. In this case, equation parameters are only a rough approximation of actual parameters.

8.9 Using capacitance values to guide interpretation of parameters obtained from EIS spectra

Corrosion, coating, and diffusion processes are not always associated with the same frequency ranges. For example, corrosion resistances are typically observed at low frequencies, but coating pore resistances can also be observed at low frequencies, particularly when a (coated metal) coating is saturated with electrolyte and metallic corrosion does not occur (referred to as capacitive dielectric corrosion behavior).[20,21]

Capacitance values can be used to guide interpretation as to what type of process is associated with each time constant. Corrosion time constants typically have capacitance values on the order of 1 to 20 microFarads per square centimeter;[22] coating time constants typically have capacitance values on the order of nanoFarads per square centimeter;[23] and oxides typically have capacitance values on the order of 1000 microFarads per square centimeter.[1] Capacitance values on the order of $\geq 100\ \mu F/cm^2$ are also believed to occur when surface adsorption occurs in conjunction with corrosion.[4]

The low frequency time constants in Figures 8.1 and 8.13 can be used to illustrate how capacitance magnitudes are used to determine what type of process is associated with a given time constant. Capacitance for the Figure 8.1 low frequency time constant is on the order of $\mu F/cm^2$, but capacitance for the Figure 8.13 low frequency time constant is on the order of 0.01 $Farad/cm^2$. Consequently, it is reasonable to conclude that the Figure 8.1 low frequency time constant is due to corrosion and the Figure 8.13 low frequency time constant is not due to corrosion, but is probably due to another process, like diffusion.

8.10 Using parameters to estimate corrosion rates and service lifetime

Corrosion rates can be calculated from EIS corrosion resistance values using the Stern Geary equation to calculate corrosion current (i_{corr} in Equation 4.2 on page 48), and Equation 4.3 to convert corrosion current to a rate (page 48).

For example, corrosion resistance is approximately 160,000 $ohms\text{-}cm^2$ in Figure 8.1. Substituting this resistance into Equation 4.2 and assuming magnitudes for Tafel slopes β_a and β_c, such as 20 mV/decade and 100 mV/decade, respectively, gives a corrosion resistance of:

$$i_{corr} \cong \frac{(20)(100)}{(2.303)(160000)(120)}$$

$$\cong 4.52\text{x}10^{-5}$$

Corrosion current is converted to a corrosion rate (MPY) using equation 4.3:

$$MPY \cong (4.52\text{x}10^{-5})(1.2866\text{x}10^{5})(1/7.86)(27.56)$$

$$\cong 20.4 \text{ MPY}$$

It is often necessary to assume or estimate Tafel slope magnitudes when calculating corrosion rates from EIS data, as was done with this example calculation. Remember from the discussion on page 50 of Chapter 4, that errors can occur from assuming or estimating Tafel slope magnitudes, causing errors in the estimated corrosion rate.

Service lifetime (for corrosion) can be estimated by dividing metal thickness by corrosion rate.

8.11 Summary

Parameters that characterize a metal (or coated metal) corrosion behavior, in an electrolyte, are obtained by regressing mathematical equations around EIS data. Statistical tests, like Chi square statistic and analysis of residuals, should be applied to regression equations, to assess how well a model (equation) actually fits experimental data. There are several computer programs that regress models around EIS data, making it easy for users to try several models to obtain the best fit to data, and thus the best estimate for corrosion and coating parameters. Capacitance magnitude can be used to help interpret what type of process is associated with a given time constant.

This chapter provided several examples of multiple time constant spectra, and time constants produced by processes other than corrosion and coatings. It was not in the scope of this book to discuss examples of every type of EIS spectrum that you may encounter. However, examples in this chapter and fundamentals in Chapter 7 should provide the knowledge needed to be able to work-out interpretations for more complex spectra and to continue learning more about how to use electrochemical impedance spectroscopy to measure corrosion.

8.12 References

1) W. S. Tait, K. A. Handrich, S. W. Tait and J. W. Martin, ASTM **STP 1188**, pp. 428-437 J. R. Scully, D. Silverman, and M. W. Kendig, Eds.,American Society for Testing and Materials, Philadelphia, PA., (1993)
2) F. Mansfeld, Corrosion, **44** (8), pp. 558-559 (1988)
3) H. Kaesche, Metallic Corrosion, p. 90, NACE, Houston, TX (1985)
4) J. N. Murray, P. J. Moran and E. Gileadi, Corrosion, **44** (8), p. 533 (1988)
5) M. Sluyters-Rehbach and J. H. Sluyters, Electroanalytical Chemistry: A Series of Advances, **4**, pp. 16-17, A. J. Bard, Ed., Marcel Dekker Inc., NY (1970)
6) R. D. Armstrong, M. F. Bell, and A. A. Metcalfe, Electrochemistry, **6**, pp. 98-127, Chemical Society Specialist Periodical Reports, UK (1978)
7) Technical Note 201, EG&G Princeton Applied Research
8) S. Chechirlian, P. Eichner, M. Keddam, H. Takenouti and H. Mazille, Electrochimica Acta, **35** (7), pp. 1125 - 1131 (1990)
9) M. Kendig and F. Mansfeld, Corrosion, **39**, p. 466 (1983)
10) B. A. Boukamp, Solid State Ionics, **18&19**, pp. 136-140 (1986)
11) J. R. Macdonald, J. Electroanal. Chem., **223**, pp. 25-50 (1987)
12) G. Walters, 2nd International Symposium on Electrochemical Impedance Spectroscopy, July 12-17, 1992, Santa Barbara, CA.
13) N. Draper and H. Smith Jr., Applied Regression Analysis, 2nd edition, p. 141 - 146, John Wiley & Sons, New York, NY (1981)
14) N. Draper and H. Smith Jr., Op. Cit., p. 4
15) I. Miller and J. E. Freund, Probability and Statistics for Engineers, third ed., p. 327 and 328, Prentice-Hall, Inc., Englewood Cliffs, NJ (1985)
16) I. Miller and J. E. Freund, Op. Cit., p. 263
17) S. Siegel and N. John Castellan, Jr., Nonparametric Statistics, McGraw-Hill Book Company, p. 45, New York, NY (1988)
18) N. Draper and H. Smith Jr., Op. Cit., p. 16
19) N. Draper and H. Smith Jr., Op. Cit., p. 148
20) W. S. Tait and J. A. Maier, J. Coating Technol., **62** (781), pp. 41-44
21) W. S. Tait, J. Coating Technol., **66** (834), pp. 59 - 61, (1994)
22) A. J. Bard and L. R. Falkner, Electrochemical Methods: Fundamentals and Applications, p.328, John Wiley & Sons, New York, NY (1980)
23) W. S. Tait, J. Coating Technol., **61** (768), pp. 57-61 (1989)

Appendix: Major corrosion mathematical equations

A. Butler Volmer equation

The Butler Volmer equation demonstrates how an electrode can become essentially anodic or cathodic when the applied potential is significantly greater than OCP (≥ 50 mV). It can also be used to calculate corrosion current density from Tafel plot data.

$$i = i_{corr}[\exp\text{-}(\eta nF\alpha/RT) - \exp((1-\alpha)nF\eta/RT)] \quad [1.11]$$

Where:

- R is the ideal gas constant, 1.986 calories/mole ^{o}K
- T is temperature in degrees Kelvin
- n is the number of electrons in the electrochemical equation
- F is Faraday's constant, 96,500 coulombs/equivalent or 23,060 Volts/equivalent
- i_{corr} is the corrosion current density in amps/cm^2
- η is the overpotential of the test electrode ($V_{applied}$ - OCP)
- α is a coefficient having values that range from 0 to 1

The anodic portion of the *BV* equation is $\exp\text{-}(\eta nF\alpha/RT)$, and its cathodic portion is $\exp((1-\alpha)nF\eta/RT)$.

B. Nernst equation

The Nernst equation is used to calculate the change in measured potential that occurs when the chemical composition of the electrical double layer changes.

$$E = E^{o} - (RT/nF)\ln[(\gamma_p[\text{products}])/(\gamma_r[\text{reactants}])] \quad [1.5]$$

Where:

- **[products]** is the concentration (in moles/liter) of chemical species on right side of an electrochemical equation
- **[reactants]** is the concentration (in moles/liter) of chemical species on the left side of the electrochemical equation,
- γ_p represents the activity coefficient for products
- γ_r represents the activity coefficient for reactants.
- **E** is the measured potential in volts or millivolts,

- **E^o** is the potential when all of the activities in Equation. 1.6 are equal to 1,
- **R** is the gas constant which is equal to 1.986 calories/mole ^oK,
- **T** is the temperature in degrees Kelvin(^oK),
- **n** is the number of electrons in the anodic electrochemical equation,
- **F** is Faraday's constant which is equal to 96,500 coulombs/equivalent or 23,060 Volts/equivalent.

The quantity RT/F is equal to 25.6 mV·equivalents at 298 ^oK (25^oC).

C. Stern Geary equation

The Stern Geary equation is used to calculate corrosion current density from linear polarization data.

$$i_{corr} = [1/(2.303R_p)][(\beta_a \cdot \beta_c)/(\beta_a+\beta_c)] \qquad [4.2]$$

Where:

- i_{corr} is the corrosion current density in Amps/cm^2,
- R_p is the corrosion resistance in ohms·cm^2,
- β_a is the anodic Tafel slope (or constant) in Volts/decade, and
- β_c is the cathodic Tafel slope (or constant) in Volts/decade

Values for the Tafel constants (slopes) are either measured (we will discuss how to measure them in the next chapter) or obtained from the corrosion literature. Values ranging from 1 mV/decade to 105 mV/decade (1000 mV = 1 Volt) have been reported as being typical.

When using the Stern Geary equation to calculate corrosion rates, it is important to remember the assumptions used for its derivation:

1) The corrosion reactions are reversible
2) Both the anodic and cathodic corrosion reactions are controlled by the activation energy for the corrosion reaction
3) Electrode surface changes do not occur during polarization
4) Polarization is due to corrosion
5) The energy barrier for the forward and reverse reactions are symmetrical

D. <u>Equation to convert corrosion current density to corrosion rate</u>

This equation converts amps/cm2 to mils (penetration) per year. One thousand mils equals one inch.

$$MPY = i_{corr}(\Lambda)(1/\rho)(\varepsilon) \qquad [4.3]$$

Where:

- Λ is $1.2866X10^5$ [equivalents•sec•mills]/[Coulombs•cm•years]
- i_{corr} is the corrosion current density in amps/cm^2 (1 Amp = 1 Coulomb/sec)
- ρ is the metal density in grams/cc, and
- ε is the equivalent weight of the metal in grams.

GLOSSARY

Activation control

A corrosion rate is activation controlled when the rate is determined by how fast a metal electrode is able to transfer its electrons to electrolyte electrochemically active species (EAS).

Active corrosion behavior:

Active corrosion behavior is typically observed when a metal produces visible quantities of corrosion after brief exposure to an electrolyte (e.g., twenty four hours exposure). Visible corrosion is typically a porous hydroxide layer that adheres loosely to the metal and does not provide very good corrosion protection. General and pitting corrosion often occur together when a metal exhibits active corrosion behavior.

Activity:

Activity is a proportionality constant used to resolve the difference between theoretical equations and experimental data.

Alternating current voltage:

An alternating current voltage is a voltage whose magnitude and polarity varies with time. The magnitude variation typically follows a sinusoidal function.

Anode:

An electrode or discrete area on an electrode surface where the oxidation of a metal (anodic half reaction) occurs.

Anodic current:

Anodic current refers to the electrical current withdrawn from a test electrode during anodic (oxidation) polarization.

Anodic half reaction:

A chemical equation that depicts the oxidation of a metal to form metal ions and free electrons. An example equation is that for iron (Fe):

$$Fe^{o} + \longrightarrow Fe^{+2} + 2e^{-}$$

Applied Voltage:

Applied voltage is a voltage (either AC or DC) that is electronically imposed on an electrode from an external voltage source such as a potentiostat.

Capacitance:

Capacitance is the ration of the amount of charge transferred between two electrodes to the voltage difference between the two electrodes. Mathematically capacitance is:

$$\text{Capacitance} = \frac{\text{Amount of Charge}}{\text{Voltage difference}}$$

Capacitive reactance

Capacitive reactance is the ability of an electrified interface (e.g., an electrical double layer) to charge (and discharge) like an electrical capacitor in response to an external applied voltage.

Cathode:

An electrode or discrete area on an electrode surface where the reduction (cathodic half reaction) of an electrochemically reactive species from the electrolyte occurs.

Cathodic current:

Cathodic current refers to the electrical current supplied to a test electrode during cathodic (reduction) polarization.

Cathodic half reaction:

A chemical equation that depicts the reduction of electrochemically active species from the electrolyte with the free electrons produced in the anodic half reaction, to form negative ions. An example equation for a cathodic half reaction is the reduction of oxygen (O_2) and water (H_2O):

$$O_2 + H_2O + 4e^- \longrightarrow 4OH^-$$

Complex Impedance (commonly referred to as Impedance)

Impedance is the AC analogue of DC resistance. Impedance magnitude is equal to AC voltage magnitude divided by the corresponding AC current magnitude.

Corrosion current density

Corrosion current density is a corrosion rate expressed as an electric current. Corrosion current densities typically have units of Amps/cm^2, but can also have units like microAmps/cm^2 or nanoAmps/cm^2.

Corrosion rate:

Corrosion rate is the amount of metal atoms converted to metal ions per unit of time. Corrosion rates can be expressed as the weight of metal lost per unit of time, or as the rate of penetration per unit of time through the metal.

Corrosion resistance:

Corrosion resistance can be defined as a metal's ability to resist corrosion, or the resistance of a metal to transfer its electrons to electrochemically active species in solution. Corrosion resistance is also called polarization resistance or charge transfer resistance and has units of ohms•cm^2 or ohms.

Counter electrode:

(see definition for electrodes)

Current Path:

A current path is the path an ion follows as it moves under the influence of an electrical field between two electrodes in an electrolyte.

Diffusion

Diffusion is ion and molecule movement caused by concentration differences between different locations in an electrolyte. Diffusion direction is from higher to lower concentrations.

Diffusion control

A corrosion rate is diffusion controlled when the rate is determined by electrolyte EAS diffusion rate to an electrode surface.

Diffusion layer:

A diffusion layer is the thin electrolyte layer near an electrode surface, whose chemical composition is different from that for the bulk electrolyte.

Ion and electrochemically active species concentrations in the diffusion layer vary from a limited number of ions and molecules on the electrode surface, to concentrations for that in the bulk electrolyte at some distance from the electrode surface. The point from the electrode surface where concentrations are equal to bulk electrolyte concentrations defines the outer edge of the diffusion layer.

Diffusion layers can be approximately 0.5 to ten millimeters thick for stagnant electrolytes,[2] or a few microns thick when electrolytes move past an electrode surface. A diffusion layer often encompasses the electrode electrical double layer.

Electrical current:

Electrical current is the flow of electrons through electrically conductive solid materials such as metal electrodes and their corresponding electrical connections to a potentiostat.

Electrical current is measured in amps, which is the amount of electrical charge in Coulombs per second. Current can be converted to a metal corrosion rate, such as milligrams per year, using Faraday's law.

EIS Spectrum

An EIS spectrum is a series of AC voltage frequencies applied to a test electrode. Typical EIS spectra for corrosion measurements include frequencies from 100 kilohertz (100,000 hertz) to several millihertz (1 millihertz = 0.001 hertz).

Electrochemically active species:

Electrochemically active species are ions or molecules (e.g., hydrogen ions or oxygen) that can be reduced by electrons.

Electrode:

An electrode is a metal submerged in an electrolyte. There are three types of electrodes used in electrochemical corrosion measurements:

Counter electrode:

The electrode used in an electrochemical corrosion test to provide the surface for the opposite half reaction (anodic or cathodic) of the test electrode.

Reference electrode:

A reference electrode is a cell containing a metal submerged in a specific concentration of its ions, plus ions from an inert salt. The cell is typically a glass tube having a semipermeable membrane or porous plug on one end that allows inert salt ions to enter and exit the cell. A reference electrode is used to measure potential difference between itself and a test electrode.

Test electrode:

A metal sample (can have any geometrical shape) such as 1018 mild steel that is submerged in an electrolyte for purposes of conducting an electrochemical corrosion test. The results from the electrochemical test are used to model the corrosion behavior of a metal, such as concrete reinforcing bars, or a metal structure, such as a steel tank.

Electrolyte:

An electrolyte (in this book) is water or aqueous solutions containing dissolved ions and/or gases, such as oxygen.

Electrochemically active species:

Electrochemically active species are ions or gases dissolved in an electrolyte that can react with electrons produced when a metal corrodes. Examples of electrochemically reactive species are hydrogen ions (H^+) or dissolved oxygen (O_2).

Electrical double layer

The electrical double layer is the electrically charged interface that occurs when a metal is submerged in an electrolyte.

Hysteresis:

Hysteresis describes the current density deviation of a cyclic polarization curve anodic back scan from the original forward anodic scan.

Negative hysteresis:

Negative hysteresis is observed when the current density of the anodic back scan is less than the current density of the original forward anodic scan.

Positive hysteresis:

Positive hysteresis is observed when the current density of the anodic back scan is larger than the current density of the original forward anodic scan.

Induction

The magnetic field from AC electrical current produces a voltage that resists making changes in current magnitude, much like friction resists motion.

Ionic current:

Ionic current is the flow of positive and negative ions under the influence of an electrical field between two electrodes in an electrolyte.

Ion mobility:

Ion mobility is the limiting velocity of an ion in a one volt electric field. Ion limiting velocity in an electrolyte is a function of ion concentration, the magnitude of charge on the ion, and frictional drag on the ion as it moves through the electrolyte.

Ion mobility is different than diffusion. Ion mobility occurs because of an electrical field between two electrodes and diffusion occurs because of a concentration difference between two areas in an electrolyte.

Mils per year (mpy)

Mils per year is the penetration rate of corrosion through a metal. One mil is 0.001 inches.

There are other corrosion rate terms, such as milligrams per square decimeter per day, but mils per year is the most common term found in corrosion journals.

Negative hysteresis:

(see the definition for hysteresis)

Open Circuit Potential:

Open circuit potential (OCP) is the electrical potential difference between two metals (submerged in an electrolyte) when no electrical current flows between them.

Overpotential:
The difference between the OCP and test electrode applied potential (voltage).

Parameter
A parameter is a single number used to characterize a data set (e.g., statistical mean), corrosion or coating behavior. Corrosion resistance and coating capacitance are examples of parameters used to characterize corrosion and coating performance, respectively.

Passive corrosion behavior:
Passive corrosion behavior occurs when a thin protective film forms on a metal surface and corrosion either does not occur, or the corrosion rate is so low that it does not significantly reduce service lifetime. The exact structure of a passive film is unknown, but it is hypothesized that passive films are 20 to 100 Angstrom thick metal oxides (1 Angstrom is 10^{-8} cm).

Polarity
Polarity is the direction of a test electrode polarization produced by an external voltage. Test electrode polarity can be either anodic, or cathodic.

Polarization:
Polarization occurs when an electrical current shifts an electrode potential from OCP.

Positive hysteresis:
(see the definition for hysteresis)

Potential
Potential is the electrical voltage difference between two electrodes, typically a reference and test electrodes. Potential is used interchangeably with voltage throughout this book.

Potentiostat:
A potentiostat is an electronic device that is used to control the potential of a test electrode in an electrolyte. The magnitude of electrode potential change (polarization) is determined by the amount of electrical current supplied by a potentiostat.
Electrode potential is measured as the difference between itself and a reference electrode (see Chapter 1 definitions on page 3).

Resonance

A parallel resistor-capacitor electrical circuit is in resonance with a given AC voltage frequency when the circuit has its maximum impedance and phase angle. Resonant frequency for a parallel resistor-capacitor circuit is proportional to circuit capacitance (in Farads) multiplied by parallel resistance (in ohms).

Scan rate

Scan rate is rate at which a potentiostat changes a test electrode potential. Scan rate has units of millivolts per second or Volts per second.

Service Lifetime

Service lifetime is the length of time prior to failure that a metallic system is exposed to an electrolyte.

Service lifetime is determined by a number of factors such as 1) corrosion rate, 2) metal thickness and 3) what the user considers to be system failure. For example, a six mils per year (mpy) pitting rate may be negligible for a vessel having 1/4 inch (250 mils) wall thickness, but the same corrosion rate would be significant for a metal container having 8 mils wall thickness. Also, pin-hole leakage may or may not be considered a failure of the vessel, but would be a metal container failure. The vessel in this example would have approximately forty two years service lifetime (250 mils/6 mpy = 41.7 years) if pin-hole leakage is considered failure, but the container service lifetime would be slightly over one year (8 mils/6 mpy = 1.3 years).

Service lifetime can also be the time it takes for corrosion to reduce metal thickness to a point where a metal structure no longer has the mechanical strength needed to continue the service for which it is designed. For example, a bridge support beam may be considered to have failed when it will no longer support its design weight (load), and its service lifetime would be the number years it took corrosion to reduce the thickness to the point where the beam can longer support its design load.

Solution resistance:

Solution resistance is the electrical resistance of an electrolyte.

Steady state

Steady state occurs when test electrode corrosion does not significantly change with time. For example, corrosion is at steady state when the corrosion rate no longer significantly changes as time increases.

Test electrode:

(see definition for electrodes)

Time constant

The time required to a) electrically charge a capacitor to a given level such as 67% maximum charge, or b) change electrical double layer (EDL) chemical composition to that needed for a given voltage magnitude, is called the circuit or EDL time constant. Time constants (in seconds) are equal to circuit or EDL capacitance (in Farads) multiplied by parallel resistance (in ohms).

SUBJECT INDEX

ERRATA

Page	**Location on page**	**Correction**
19	The first and second paragraphs in section 2.3	Equation 1.11 in the text should be Equation **1.9**
23	Under Equation [2.2]	(Equation **1.9** on page 11)
39	6th line of the 3rd paragraph	Sixteen **days**
90	Line 4 of the second paragraph	(Equation 7.1 on page **83**)
96	First paragraph in section 8.3	Chapter 7 Figures 7.8 through 7.10 (pages **90 and** 91)
108	List of software programs	**Scribner Associates, Inc. distributed by Solartron, Inc.**
109	Line 4 of the second paragraph	… on pages **83 and 85,** respectively
110	Figure 8.21	$\mathbf{R^{**} = R_{\Omega} + R_{coat}}$
117	Butler Volmer Equation	The equation number is **[1.9]**
125	Electrochemically Active Species	The two definitions should be combined